AutoCAD 2019 For Novices

(Learn By Doing)

CADSoft Technologies

AutoCAD 2019 for Novices: Learn By Doing

CADSoft Technologies

CADSoft Technologies

ISBN: 978-1-64057-045-0

NOTICE TO THE READER

Publisher does not warrant or guarantee any of the products described in the text or perform any independent analysis in connection with any of the product information contained in the text. Publisher does not assume, and expressly disclaims,any obligation to obtain and include information other than that provided to it by the manufacturer.

The reader is expressly warned to consider and adopt all safety precautions that might be indicated by the activities herein and to avoid all potential hazards. By following the instructions contained herein, the reader willingly assumes all risks in connection with such instructions.

The Publisher makes no representation or warranties of any kind, including but not limited to, the warranties of fitness for particular purpose or merchantability, nor are any such representations implied with respect to the material set forth herein, and the publisher takes no responsibility with respect to such material. The publisher shall not be liable for any special, consequential, or exemplary damages resulting, in whole or part, from the reader's use of, or reliance upon, this material.

DEDICATION

To students, who are dedicated to learning new technologies

THANKS

To the faculty and employees of
CADSoft Technologies for their cooperation

Training Programs offered by CADSoft Technologies

CADSoft Technologies provides quality training on various software packages including Computer Aided Design, Manufacturing and Engineering (CAD/CAM/CAE), Web and Programming Languages, animation, architecture, and GIS. The training is delivered at its training centers or online at any time, place, and pace to individuals, corporates as well as the students of colleges, universities, and CAD/CAM/CAE training centers. The main features of this program are:

Training in Classroom Settings and Online Training

A team of highly experienced instructors and qualified engineers at CADSoft Technologies conduct the classes in classroom environment at its centers. This team has authored several textbooks that are rated "one of the best" in their categories and are used in various colleges, universities, and training centers in North America, Europe, and in other parts of the world.

CADSoft Technologies with its cost effective and time saving initiative strives to deliver online training in the comfort of your home or work place, thereby relieving you from the hassles of traveling to training centers. CADSoft Technologies strives to be the best training institute in USA. CADSoft Technologies has developed career oriented programs by following state-of-the-art teaching and learning methodologies.

Courses Offered by CADSoft Technologies

CADSoft Technologies conducts classes on the following software packages:

CAD/CAM/CAE: *CATIA, Pro/ENGINEER Wildfire, Creo Parametric, Creo Direct, SOLIDWORKS, Autodesk Inventor, Solid Edge, NX, AutoCAD, AutoCAD LT, AutoCAD Plant 3D, Customizing AutoCAD, SolidCAM, NX CAM, NX Mold, Creo Mold, Alias Design, Alias Automotive, Ansys Fluent, and ANSYS*

Architecture and GIS: *Autodesk Revit Architecture, AutoCAD Civil 3D, Autodesk Revit Structure, AutoCAD Map 3D, Revit MEP, Navisworks, Primavera, and Bentley STAAD Pro, MS Project, MX Road, ArcGIS, and Raster Design*

Animation and VFX: *Autodesk 3ds Max, Chaosgroup VRay, Autodesk 3ds Max Design, Autodesk Maya, Blackmagic Design Fusion Studio, Adobe Premiere, Adobe Photoshop, Adobe Indesign, Adobe Illustrator, Corel Graphic Design CorelDraw, The Foundry NukeX, and MAXON CINEMA 4D*

Web and Programming: *C/C++, HTML5/CSS3, JavaScript, jQuery, Bootstrap, PHP, MySQL, Dreamweaver, VB.NET, Oracle, AJAX, and Java*

Personality Development: *Personality Development and Engineering Ethics/Soft Skills Course*

Table of Contents

Chapter 4: Working with Drawing Aids

Chapter 5: Editing Sketched Objects-I

Chapter 6: Basic Dimensioning, Geometric Dimensioning, and Tolerancing

Chapter 9: Creating Texts and Tables

Chapter 10: Editing Sketched Objects-II

Chapter 11: Adding Constraints to Sketches

Chapter 12: Hatching Drawings

Chapter 13: Model Space Viewports, Paper Space Viewports, and Layouts

Chapter 19: Working with Advanced Drawing Options

Chapter 20: Grouping and Advanced Editing of Sketched Objects

Chapter 21: Working with Data Exchange & Object linking and Embedding

This page is intentionally left blank

Preface

AutoCAD 2019

AutoCAD, developed by Autodesk Inc., is the most popular PC-CAD system available in the market. Today, over 7 million people use AutoCAD and other AutoCAD-based design products. 100% of the Fortune 100 firms and 98% of the Fortune 500 firms are Autodesk customers. AutoCAD's open architecture allows third-party developers to write application software which has significantly added to its popularity. For example, the author of this Textbook has developed a software package "**SMLayout**" for sheet metal products that generates a flat layout of various geometrical shapes such as transitions, intersections, cones, elbows, tank heads, and so on. Several companies in Canada and United States are using this software package with AutoCAD to design and manufacture various products. AutoCAD also facilitates customization that enables the users to increase their efficiency and improve their productivity.

The **AutoCAD 2019 for Novices: Learn By Doing** has been created to explain the basic concepts and tools of AutoCAD. This book comprises of solved tutorials and exercises which will enable users to learn and practice concepts of AutoCAD.

Some of the main features of the textbook are as follows:

- **Real-world Mechanical Engineering Projects as Tutorials**
  The author has used the real-world mechanical engineering projects as tutorials in this textbook so that the readers can correlate the tutorials with the real-time models in the mechanical engineering industry.

- **Tips and Notes**
  Additional information related to various topics is provided to the users in the form of tips and notes.

Symbols Used in the Textbook

Note

The author has provided additional information to the users about the topic being discussed in the form of notes.

Tip

Special information and techniques are provided in the form of tips that will increase the efficiency of the users.

Formatting Conventions Used in the Textbook

Refer to the following list for the formatting conventions used in this textbook.

- Command names are capitalized and written in boldface letters.

  Example: The **MOVE** command

- A key icon appears when you have to respond by pressing the ENTER or the RETURN key.

- Command sequences are indented. The responses are indicated in boldface. The directions are indicated in italics and the comments are enclosed in parentheses.

  Command: **MOVE**
  Select object: **G**
  Enter group name: *Enter a group name (the group name is group1)*

- The methods of invoking a tool/option from the **Ribbon**, **Menu Bar**, **Quick Access Toolbar**, **Tool Palettes**, **Application menu**, toolbars, Status Bar, and Command prompt are enclosed in a shaded box.

Ribbon:	Draw > Line
Menu Bar:	Draw > Line
Tool Palettes:	Draw > Line
Toolbar:	Draw > Line
Command:	LINE or L

Naming Conventions Used in the Textbook

Tool

If you click on an item in a toolbar or a panel of the **Ribbon** and a command is invoked to create/edit an object or perform some action, then that item is termed as **tool**.

For example:
To Create: **Line** tool, **Circle** tool, **Extrude** tool
To Edit: **Fillet** tool, **Array** tool, **Stretch** tool
Action: **Zoom** tool, **Move** tool, **Copy** tool

If you click on an item in a toolbar or a panel of the **Ribbon** and a dialog box is invoked wherein you can set the properties to create/edit an object, then that item is also termed as **tool**, refer to Figure 1.

For example:
To Create: **Define Attributes** tool, **Create** tool, **Insert** tool
To Edit: **Edit Attributes** tool, **Block Editor** tool

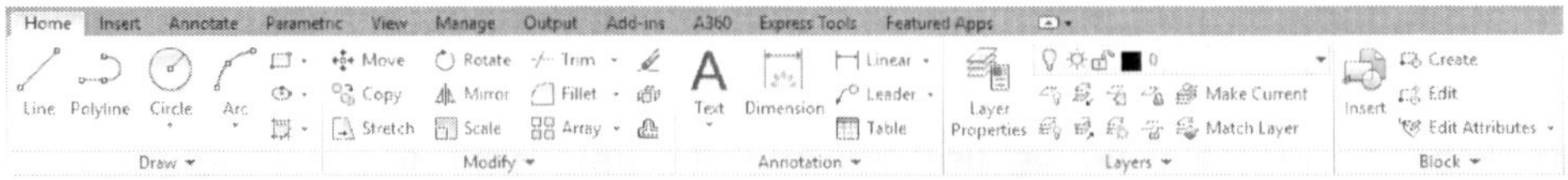

Figure 1** Various tools in the **Ribbon

Button

If you click on an item in a toolbar or a panel of the **Ribbon** and the display of the corresponding object is toggled on/off, then that item is termed as **button**. For example, **Grid** button, **Snap Mode** button, **Ortho Mode** button, **Properties** button, **Tool Palettes** button, and so on; refer to Figure 2.

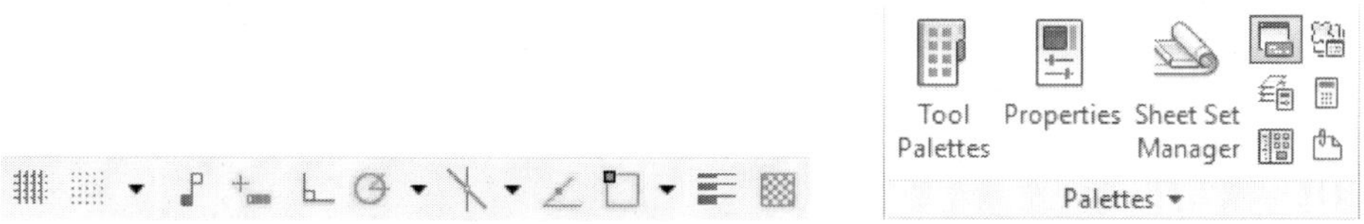

Figure 2** Various buttons displayed in the Status Bar and **Ribbon

The item in a dialog box that has a 3d shape like a button is also termed as **button**. For example, **OK** button, **Cancel** button, **Apply** button, and so on.

Dialog Box

The naming conventions used for the components in a dialog box are mentioned in Figure 3.

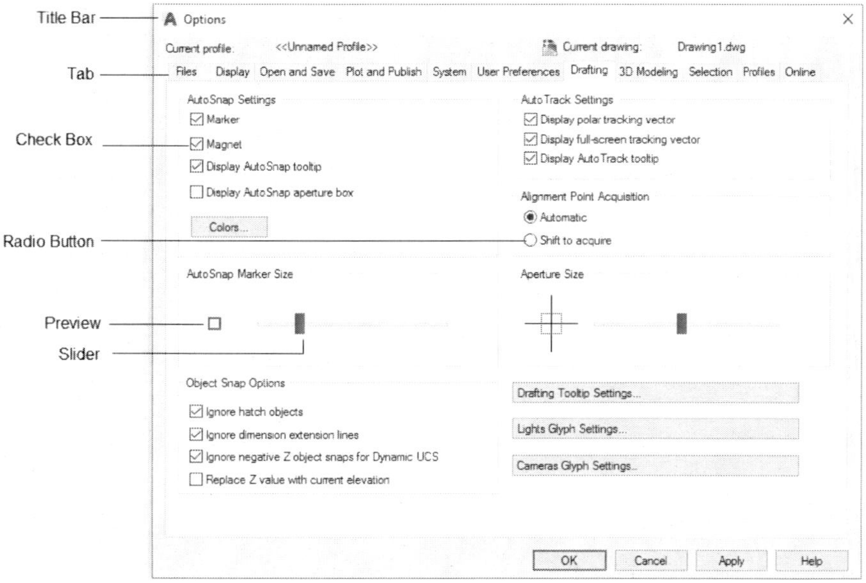

***Figure 3** The components of a dialog box*

Drop-down

A drop-down is the one in which a set of common tools are grouped together. You can identify a drop-down with a down arrow on it. These drop-downs are given a name based on the tools grouped in them. For example, **Circle** drop-down, **Fillet** drop-down, and so on; refer to Figure 4.

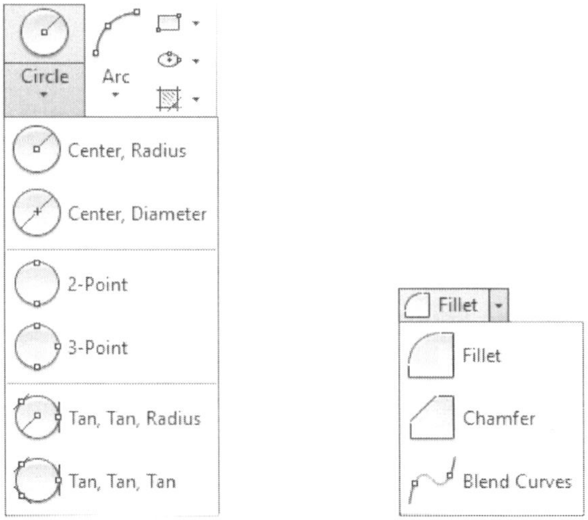

*Figure 4 The **Circle** and **Fillet** drop-downs*

Drop-down List

A drop-down list is the one in which a set of options are grouped together. You can set various parameters using these options. You can identify a drop-down list with a down arrow on it. To know the name of a drop-down list, move the cursor over it; its name will be displayed as a tool tip. For example, **Lineweight** drop-down list, **Linetype** drop-down list, **Object Color** drop-down list, **Visual Styles** drop-down list, and so on; refer to Figure 5.

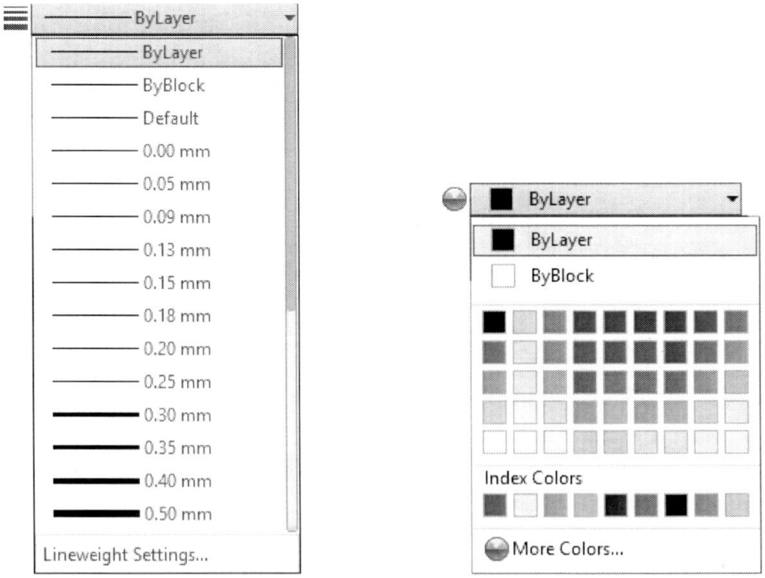

*Figure 5 The **Lineweight** and **Object Color** drop-down lists*

Options

Options are the items that are available in shortcut menu, drop-down list, Command prompt, **PROPERTIES** palette, and so on. For example, choose the **Properties** option from the shortcut menu displayed on right-clicking in the drawing area; refer to Figure 6.

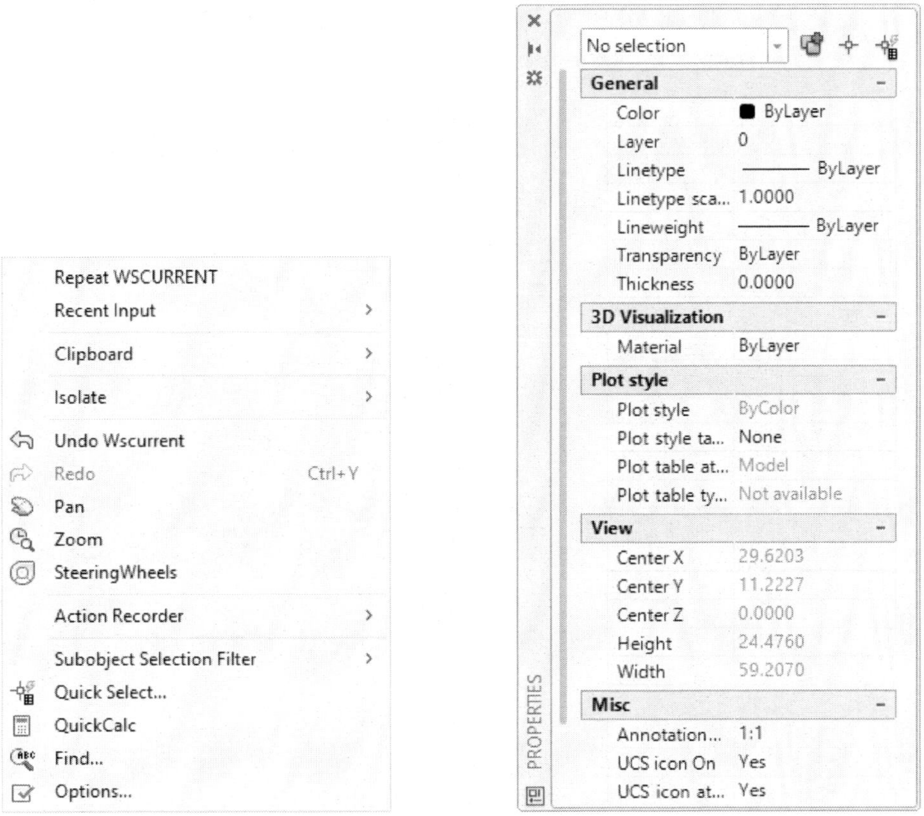

***Figure 6** Options in the shortcut menu and the **PROPERTIES** palette*

Tools and Options in Menu Bar

A menu bar consists of both tools and options. As mentioned earlier, the term **tool** is used to create/edit something or to perform some action. For example, in Figure 7, the item **Box** has been used to create a box shaped surface, therefore it will be referred to as the **Box** tool.

Similarly, an option in the menu bar is the one that is used to set some parameters. For example, in Figure 7, the item **Linetype** has been used to set/load the linetype, therefore it will be referred to as an option.

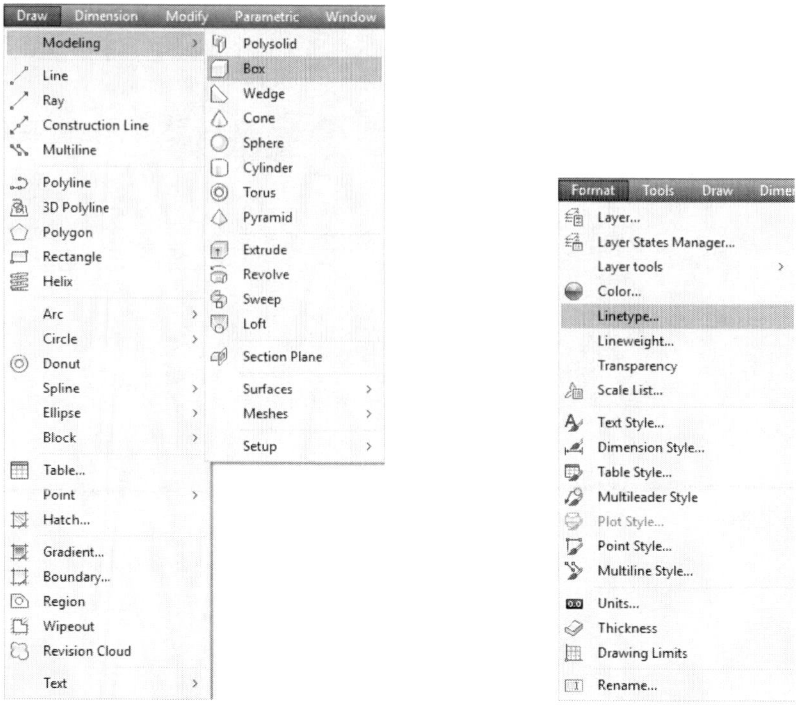

Figure 7 *Tools and options in the menu bar*

Chapter 1

Introduction to AutoCAD

Learning Objectives

After completing this chapter, you will be able to:

- *Start AutoCAD*
- *Invoke AutoCAD commands from the keyboard, menus, toolbars, shortcut menus, Tool Palettes, and Ribbon*
- *Start a new drawing using the New tool*
- *Save the work using various file-saving commands*
- *Close a drawing*
- *Open an existing drawing*
- *Quit AutoCAD*
- *Save data to Web & Mobile*

AutoCAD SCREEN COMPONENTS

When you install AutoCAD 2019, the AutoCAD 2019 - English shortcut icon is created on the desktop. You can start AutoCAD by double-clicking on this icon. The initial AutoCAD screen comprises of drawing area, command window, menu bar, several toolbars, Model and Layout tabs, and Status Bar, refer to Figure 1-1. A title bar that has an AutoCAD symbol and the current drawing name is displayed on top of the screen.

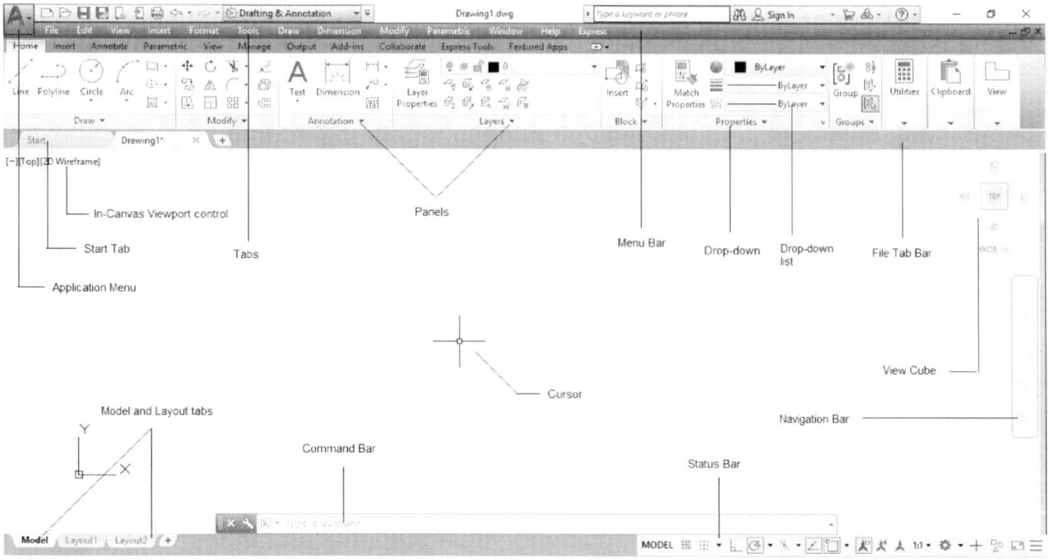

Figure 1-1 AutoCAD screen components in AutoCAD ***Drafting & Annotation*** *workspace*

Start Tab

In AutoCAD, the **Start** tab is displayed in the AutoCAD environment when you close all the drawing templates or when there are no drawings open. The **Start** tab contains two sliding frames, **CREATE** and **LEARN**, refer to Figure 1-2. These frames are discussed next.

CREATE

The **CREATE** page is displayed by default. In the **CREATE** page, you can access sample file, recent files, templates, product notifications as well as connect with the online community. The **CREATE** page is divided into four areas: **Get Started**, **Recent Documents**, **Notifications**, and **Connect**.

LEARN

When you click on **LEARN**, the **LEARN** page is displayed. The **LEARN** page provides tools to help you learn AutoCAD. It is divided into three columns: **What's New**, **Getting Started Videos**, and **Learning Tips/Online Resources**.

Drawing Area

The drawing area covers the major portion of the screen. In this area, you can draw the objects and use the commands. To draw the objects, you need to define the coordinate points, which can be selected by using your pointing device. The position of the pointing device is represented on the screen by the cursor. The window also has the standard Windows buttons such as Close, Minimize, and Restore Down on the top right corner. These buttons have the same functions as in any other standard Window.

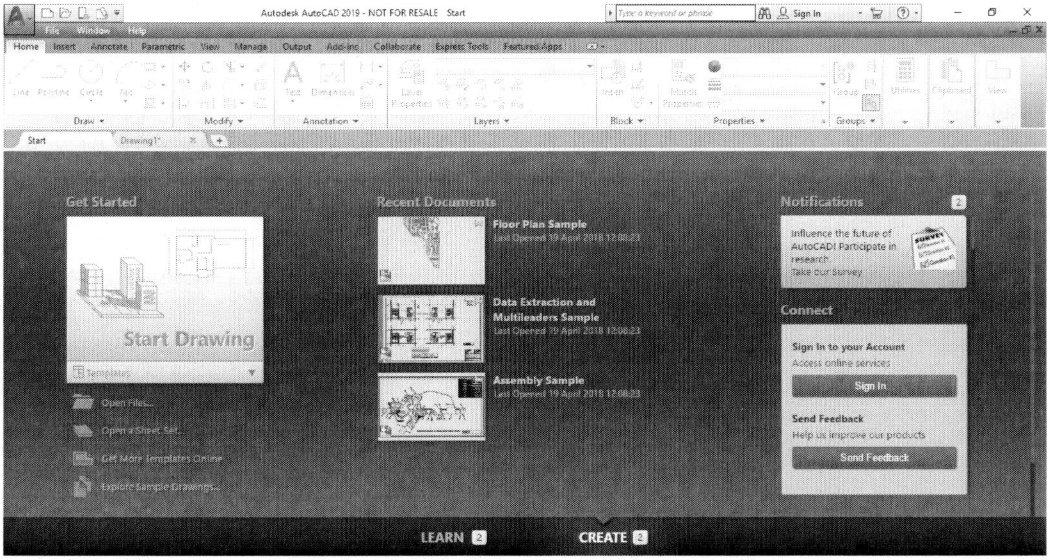

Figure 1-2 *The **Start** tab in AutoCAD 2019*

Command Window

The command window at the bottom of the drawing area has the command prompt where you can enter the commands. It also displays the subsequent prompt sequences and the messages. You can change the size of the window by placing the cursor on the top edge (double line bar known as the grab bar) and then dragging it. This way you can increase its size to see all the previous commands you have used. You can also press the F2 key to display **AutoCAD Text Window**, which displays the previous commands and prompts.

Tip

*You can hide all toolbars displayed on the screen by pressing the CTRL+0 keys or by choosing **View > Clean Screen** from the menu bar. To turn on the display of the toolbars again, press the CTRL+0 keys. You can also choose the **Clean Screen** button in the Status Bar to hide all toolbars.*

INVOKING TOOLS IN AutoCAD

On starting AutoCAD, when you are in the drawing area, you need to invoke AutoCAD tools to perform an operation. For example, to draw a line, first you need to invoke the **Line** tool and then define the start point and the endpoint of the line. Similarly, if you want to erase objects, you must invoke the **Erase** tool and then select the objects for erasing. In AutoCAD, you can invoke the commands using the Keyboard, Ribbon, Application Menu, Tool Palettes, Menu bar, Shortcut menu, and Toolbar.

STARTING A NEW DRAWING

Application Menu: New > Drawing	**Menu Bar:** File > New
Quick Access Toolbar: New	**Command:** NEW or QNEW

You can open a new drawing using the **New** tool in the **Quick Access Toolbar**. When you invoke the **New** tool, by default AutoCAD will display the **Select template** dialog box, as shown in Figure 1-3. This dialog box displays a list of default templates available in AutoCAD. The default selected template is ***acad.dwt***, which starts the 2D drawing environment. You can select the ***acad3D.dwt*** template to start the 3D modeling environment. Alternatively, you can select any other template to start a new drawing that will use the settings of the selected template. You can also open any drawing without using any template either in metric or imperial system. To do so, choose the down arrow on the right of the **Open** button and choose the **Open with no Template-Metric** option or the **Open with no Template-Imperial** option from the drop-down.

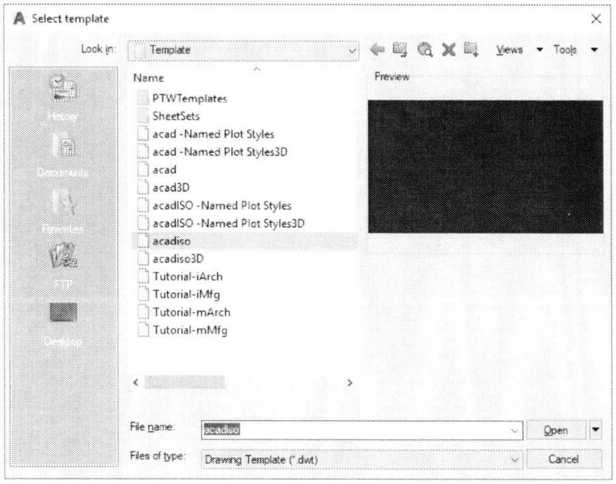

***Figure 1-3** The **Select template** dialog box*

SAVING YOUR WORK

Application Menu: Save, Save As	**Menu Bar:** File > Save or Save As
Quick Access Toolbar: Save or Save As	**Command:** QSAVE, SAVEAS, SAVE

You must save your work before you exit the drawing editor or turn off your system. Also, it is recommended that you save your drawings in regular intervals, so that in the event of a power failure or an editing error, all work saved before the problem started is retained.

AutoCAD has provided the **QSAVE**, **SAVEAS**, and **SAVE** commands that allow you to save your work. Also, these commands allow you to save your drawing by writing it to a permanent storage device, such as a hard drive or in any removable drive.

CLOSING A DRAWING

You can close the current drawing file without actually quitting AutoCAD by choosing **Close > Current Drawing** from the **Application Menu** or by entering **CLOSE** at the command prompt.

If multiple drawing files are opened, choose **Close > All Drawings** from the **Application Menu**. If you have not saved the drawing after making the last changes to it and you invoke the **CLOSE** command, AutoCAD displays a dialog box that allows you to save the drawing before closing. This box gives you an option to discard the current drawing or the changes made to it. It also gives you an option to cancel the command. After closing the drawing, you are still in AutoCAD from where you can open a new or an already saved drawing file. You can also use the close button (**X**) of the drawing area to close the drawing.

OPENING AN EXISTING DRAWING

Application Menu: Open > Drawing **Quick Access Toolbar:** Open
Menu Bar: File > Open **Command:** OPEN

You can open an existing drawing file that has been saved previously. There are three methods that can be used to open a drawing file: by using the **Select File** dialog box, by using the **Create New Drawing** dialog box, and by dragging and dropping.

QUITTING AutoCAD

You can exit the AutoCAD program by using the **EXIT** or **QUIT** commands. Even if you have an active command, you can choose the **Exit Autodesk AutoCAD 2019** from the **Application Menu** to quit the AutoCAD program. In case the drawing has not been saved, it allows you to save the work first through a dialog box. Note that if you choose **No** in this dialog box, all the changes made in the current drawing till the last save will be lost. You can also use the **Close** button (**X**) of the main AutoCAD window (present in the title bar) to end the AutoCAD session.

SAVE TO WEB & MOBILE

New

Application Menu: Save As > Drawing to AutoCAD Web & Mobile
Quick Access Toolbar: Save to Web & Mobile **Command:** SAVETOWEBMOBILE

Using the **Save to Web & Mobile** tool, you can save copy of your drawings in your Autodesk web & mobile account from any remote location in the world using any device such as desktop or mobile having internet access. You can access this tool from the **Quick Access Toolbar** or **Application Menu**. When you choose this tool for the first time, you will be prompted to install the Save to AutoCAD Web and Mobile Plug-in. Choose the **Save to Web & Mobile** tool from the **Quick Access Toolbar**; the **Install Save to AutoCAD Web & Mobile** window will be displayed, refer to Figure 1-4.

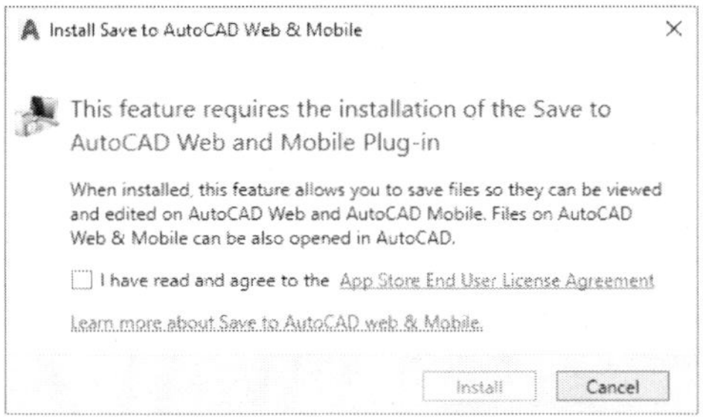

*Figure 1-4 The **Install Save to AutoCAD Web & Mobile** window*

Select the **I have read and agree to the App Store End User License Agreement** check box from this window; the **Install** button will be activated. Next, choose the **Install** button. Once the installation process is complete; the **Save in AutoCAD Web & Mobile Cloud Files** dialog box will be displayed, refer to Figure 1-5.

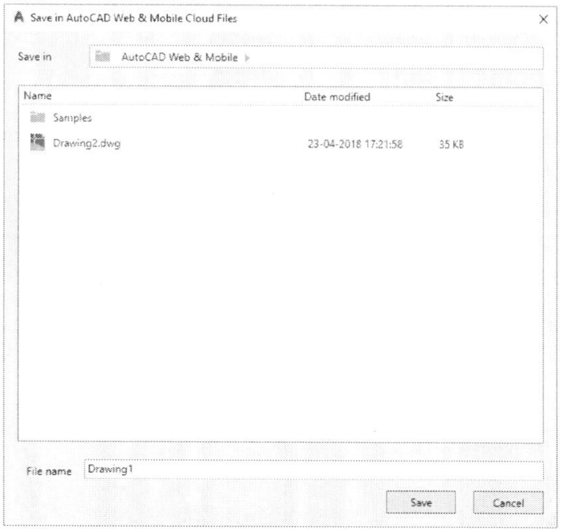

Figure 1-5 *The* ***Save in AutoCAD Web & Mobile Cloud Files*** *dialog box*

Enter the name of the file to be saved in the **File name** edit box. Next, choose the **Save** button.

Note

1. The name of the saved files appear in the ***Name*** *column of the list box available in the* ***Save in AutoCAD Web & Mobile Cloud Files*** *dialog box when you invoke this dialog box again.*

2. You need to first sign in to your Autodesk account to use this tool. If you are not signed in and choose this tool, the Autodesk sign-in window will appear and will prompt you to sign in first.

3. The drawing files saved in your web and mobile account are saved to the cloud and utilise the cloud space. You can save them to your device by using the ***Save As*** *tool.*

4. You can also share the saved files with any of the co-workers or clients across the world. They can review or edit the drawing file depending upon the permissions you grant them.

After saving the file to your web and mobile account, you can access it from anywhere across the world using any device (mobile, tablet, etc) having Wi-fi or internet connection.

Note

You can access the saved web and mobile files using the ***Open from Web & Mobile*** *tool available in the* ***Quick Access Toolbar****.*

Chapter 2

Getting Started with AutoCAD

Learning Objectives

After completing this chapter, you will be able to:

- *Draw lines by using the Line tool*
- *Understand the coordinate systems used in AutoCAD*
- *Clear the drawing area by using the Erase tool*
- *Draw circles*

DYNAMIC INPUT MODE

In AutoCAD, the Dynamic Input mode allows you to enter the commands through the pointer input and the dimensions using the dimensional input. You can turn ON/OFF the Dynamic Input mode by using the **Dynamic Input** button available in the Status Bar (Customize to Add). When you start AutoCAD for the first time, the Dynamic Input mode is active, as the **Dynamic Input** button is chosen by default in the Status Bar. With this mode turned on, all the prompts are available at the tooltip as dynamic prompts and you can select the command options through the dynamic prompt. The settings for the Dynamic Input mode are done through the **Dynamic Input** tab of the **Drafting Settings** dialog box. To invoke the **Drafting Settings** dialog box, right-click on the **Dynamic Input** button in the Status Bar; a shortcut menu will be displayed. Choose the **Dynamic Input Settings** option from the shortcut menu; the **Drafting Settings** dialog box will be displayed, as shown in Figure 2-1.

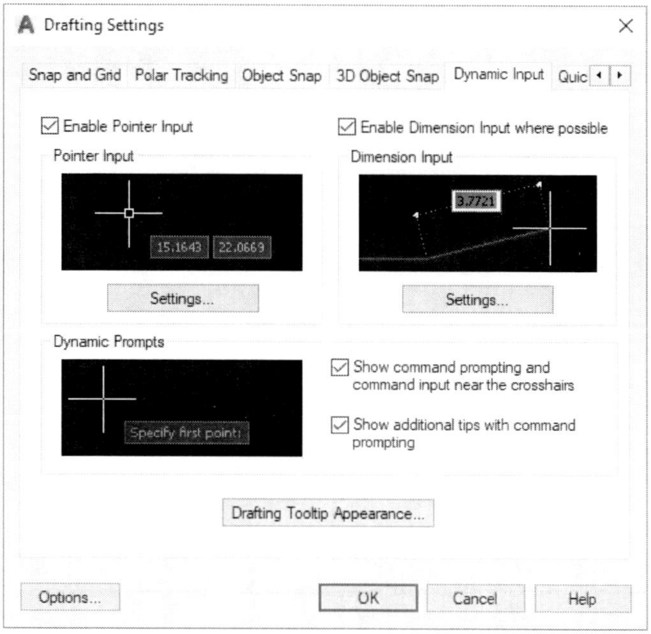

__Figure 2-1__ The __Dynamic Input__ tab of the __Drafting Settings__ dialog box

Note

If the __Dynamic Input__ button is not available in the Status Bar, then you can customize it using the __Customization__ button which is available at the right corner of the Status Bar.

DRAWING LINES IN AutoCAD

Ribbon: Home > Draw > Line **Toolbar:** Draw > Line **Menu Bar:** Draw > Line
Tool Palettes: Draw > Line **Command:** LINE/L

The most commonly used fundamental object in a drawing is line. In AutoCAD, a line is drawn between two points by using the **Line** tool. You can invoke the **Line** tool from the **Draw** panel of the **Home** tab in the **Ribbon**, refer to Figure 2-2. Besides this, you can choose the **Line** tool from the **Draw** tab of the **Tool Palettes**. To invoke the **Tool Palettes**, choose the **Tool Palettes** button from the **Palettes** panel in the **View** tab, as shown in

Figure 2-3. Alternatively, you can invoke the **Line** tool from the **Draw** toolbar, as shown in Figure 2-4. However, the **Draw** toolbar is not displayed by default. To invoke this toolbar, choose **Tools > Toolbars > AutoCAD > Draw** from the Menu Bar.

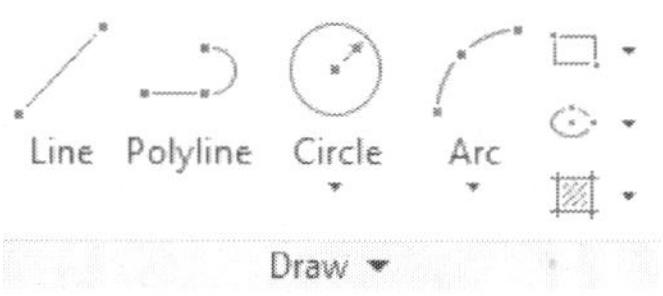

*Figure 2-2 The **Line** tool in the **Draw** panel*

*Figure 2-3 Invoking the **Tool Palettes** from the **Palettes** panel*

*Figure 2-4 The **Line** tool in the **Draw** toolbar*

You can also invoke the **Line** tool by entering **LINE** or **L** (L is the alias for the **LINE** command) at the command prompt. On invoking the **Line** tool, you will be prompted to specify the starting point of the line. Specify a point by clicking the left mouse button in the drawing area or by entering its coordinates in the Dynamic Input fields or the command prompt. After specifying the first point, you will be prompted to specify the second point. Specify the second point; a line will be drawn, refer to Figure 2-5. You may continue specifying points and draw lines or terminate the **Line** tool by pressing ENTER, ESC, or SPACEBAR. You can also right-click to display the shortcut menu and then choose the **Enter** or **Cancel** option from it to exit the **Line** tool.

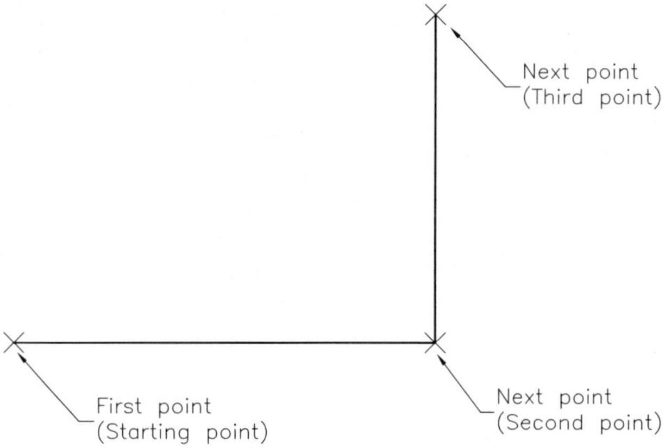

*Figure 2-5 Drawing lines using the **Line** tool*

COORDINATE SYSTEMS

In AutoCAD, the location of a point is specified in terms of Cartesian coordinates. In this system, each point in a plane is specified by a pair of numerical coordinates. To specify a point in a plane, take two mutually perpendicular lines as references. The horizontal line is called the *X* axis and the vertical line is called the *Y* axis. The *X* and *Y* axes divide the *XY* plane into four parts, generally known as quadrants. The point of intersection of these two axes is called the origin and the plane is called the XY plane. The origin has the coordinate values of X = 0,

Y = 0. The origin is taken as the reference for locating a point on the *XY* plane. Now, to locate a point, say P, draw a vertical line intersecting the X axis. The horizontal distance between the origin and the intersection point will be called the X coordinate of P. It will be denoted as P(x). The *X* coordinate specifies how far the point is to the left or right from the origin along the *X* axis. Now, draw a horizontal line intersecting the Y axis. The vertical distance between the origin and the intersection point will be the Y coordinate of P. It will be denoted as P(y). The *Y* coordinate specifies how far the point is to the top or bottom from the origin along the *Y* axis. The intersection point of the horizontal and vertical lines is the coordinate of the point and is denoted as P(x,y). The *X* coordinate is positive if measured from the right of the origin and is negative if measured from the left of the origin. The *Y* coordinate is positive if measured above the origin and is negative if measured below the origin, refer to Figure 2-6.

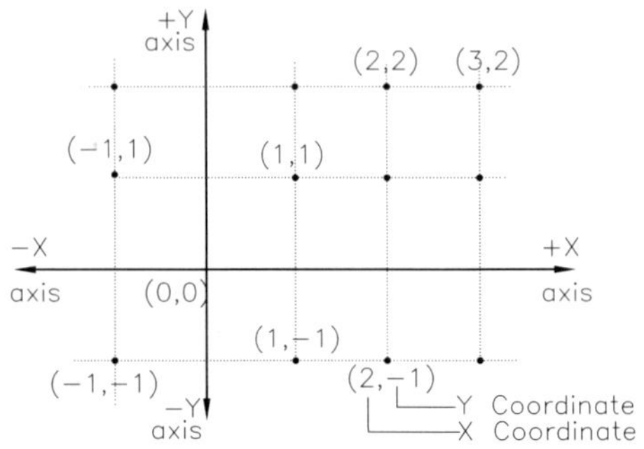

Figure 2-6 *Cartesian coordinate system*

In AutoCAD, the default origin is located at the lower left corner of the drawing area. AutoCAD uses the following coordinate systems to locate a point in the XY plane.

1. Absolute coordinate system
2. Relative coordinate system
    a. Relative rectangular coordinate system
    b. Relative polar coordinate system
3. Direct distance entry

If you are specifying a point by entering its location at the command prompt then you need to use any one of the coordinate systems.

Absolute Coordinate System

In the Absolute Coordinate System, points are located with respect to the origin (0,0). For example, a point with coordinates, X = 4 and Y = 3 is measured 4 units horizontally (distance along the *X* axis) and 3 units vertically (distance along the *Y* axis) from the origin, as shown in Figure 2-7. In AutoCAD, the absolute coordinates are specified at the command prompt by entering *X* and *Y* coordinates, separated by a comma. However, remember that if you are specifying the coordinates by using the Dynamic Input mode, you need to add # as the prefix to the X coordinate value. The following example illustrates the use of absolute coordinates at the command prompt to draw the rectangle shown in Figure 2-8.

Choose the ***Line*** *tool (Ensure that the* ***Dynamic Input*** *button is not chosen)*

LINE Specify first point: **1,1** Enter *(X = 1 and Y = 1.)*

Specify next point or [Undo]: **4,1** Enter *(X = 4 and Y = 1.)*

Specify next point or [Undo]: **4,3** Enter

Specify next point or [Close /Undo]: **1,3** Enter

Specify next point or [Close/Undo]: **C** Enter

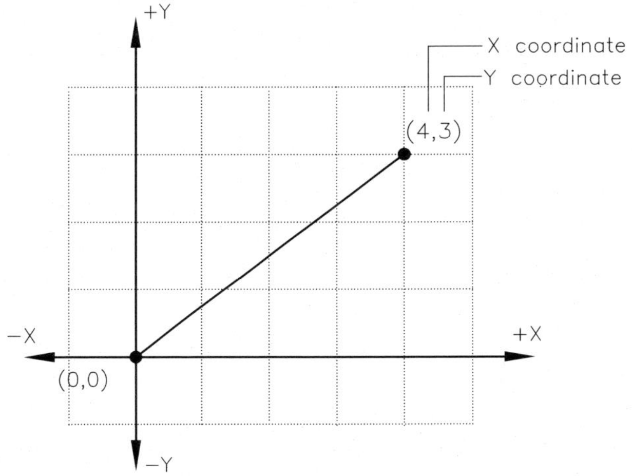

Figure 2-7 *Absolute Coordinate System*

Figure 2-8 *Rectangle created by using absolute coordinates*

Tutorial 1 Absolute Coordinate System

Draw the profile shown in Figure 2-9 by using the Absolute Coordinate system. The absolute coordinates of the points are given in the table given below. Save the drawing with the name *Tut1.dwg*.

Point	Coordinates	Point	Coordinates
1	3,1	5	5,2
2	3,6	6	6,3
3	4,6	7	7,3
4	4,2	8	7,1

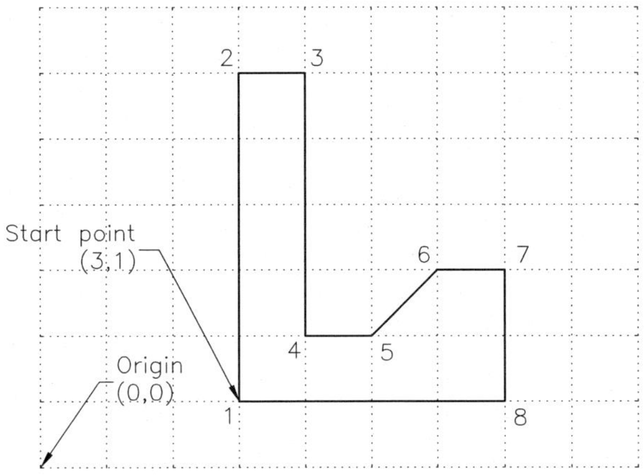

Figure 2-9 *Drawing a figure using the absolute coordinates*

Start a new file with the ***acad.dwt*** template in the **Drafting & Annotation** workspace. Once you know the coordinates of the points, you can draw the sketch by using the **Line** tool.

*Choose the **Zoom All** tool*

*Choose the **Line** tool*

The prompt sequence is given next.

LINE Specify first point: **3,1** Enter *(Start point.)*

Specify next point or [Undo]: **3,6** Enter

Specify next point or [Undo]: **4,6** Enter

Specify next point or [Close/Undo]: **4,2** Enter

Specify next point or [Close/Undo]: **5,2** Enter

Specify next point or [Close/Undo]: **6,3** Enter

Specify next point or [Close/Undo]: **7,3** Enter

Specify next point or [Close/Undo]: **7,1** Enter

Specify next point or [Close/Undo]: **C** Enter

Choose the **Save** tool from the **Quick Access Toolbar** to display the **Save Drawing As** dialog box. Enter **Tut1** in the **File name** edit box and then choose the **Save** button. The drawing will be saved with the specified name in the default *Documents* folder.

Relative Coordinate System

There are two types of relative coordinate system:
1. Relative Rectangular System
2. Relative Polar System

Tutorial 2 — Relative Rectangular Coordinate

Draw the profile shown in Figure 2-10 using Relative Rectangular Coordinates. The coordinates of the points are given in the table below.

Point	Coordinates	Point	Coordinates
1	3,1	8	-1,-1
2	4,0	9	-1,1
3	0,1	10	-1,0
4	-1,0	11	0,-2
5	1,1	12	1,-1
6	0,2	13	-1,0
7	-1,0	14	0,-1

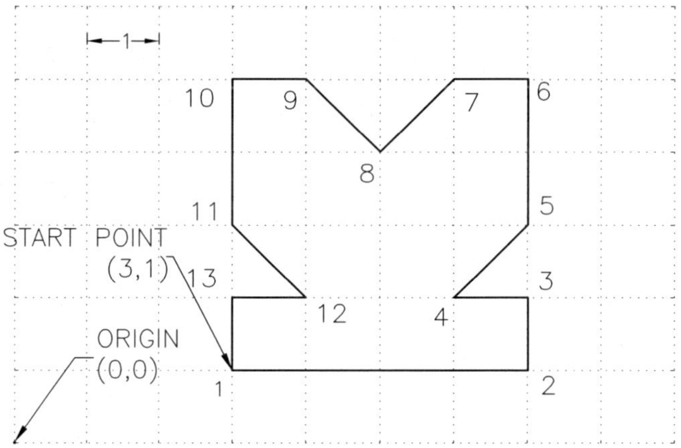

Figure 2-10 *Profile for Tutorial 2*

Start a new file with the ***acad.dwt*** template in the **Drafting & Annotation** workspace. Before you proceed, you need to make sure that the **Dynamic Input** is turned on.

Choose the ***Zoom All*** tool

Next, *choose the **Line** tool*

LINE Specify first point: *Type* **3,1** *in the dynamic input boxes and press* Enter *(Start point)*

Specify next point or [Undo]: *Type* **4,0** *in the dynamic input boxes and press* Enter

Specify next point or [Undo]: *Type* **0,1** *in the dynamic input boxes and press* Enter

Specify next point or [Close/Undo]: *Type* **-1,0** *in the dynamic input boxes and press* Enter

Specify next point or [Close/Undo]: *Type* **1,1** *in the dynamic input boxes and press* Enter

Specify next point or [Close/Undo]: *Type* **0,2** *in the dynamic input boxes and press* Enter

Specify next point or [Close/Undo]: **-1,0** *and press* Enter

Specify next point or [Close/Undo]: **-1,-1** *and press* Enter

Specify next point or [Close/Undo]: **-1,1** *and press* Enter

Specify next point or [Close/Undo]: **-1,0** *and press* Enter

Specify next point or [Close/Undo]: **0,-2** *and press* Enter

Specify next point or [Close/Undo]: **1,-1** *and press* Enter

Specify next point or [Close/Undo]: **-1,0** *and press* Enter

Specify next point or [Close/Undo]: **0,-1** *and press* Enter

Specify next point or [Close/Undo]: Enter

Direct Distance Entry

The easiest way to draw a line in AutoCAD is by using the Direct Distance Entry method. Before drawing a line by using this method, ensure that the **Dynamic Input** button is chosen in the Status Bar.

ERASING OBJECTS

Ribbon: Home > Modify > Erase **Toolbar:** Modify > Erase
Menu Bar: Modify > Erase **Tool Palettes:** Modify > Erase
Command: ERASE/E

Sometimes, you may need to erase the unwanted objects from the drawing. You can do so by using the **Erase** tool. To erase an object, choose the **Erase** tool from the **Modify** panel, refer to Figure 2-11. You can also choose the **Erase** tool from the **Modify** toolbar, as shown in Figure 2-12. To invoke the **Modify** toolbar, choose **Tools > Toolbars > AutoCAD > Modify** from the Menu Bar. On invoking the **Erase** tool, a small box, known as Pick box, replaces the screen cursor. To erase the object, select it by using the pick box, refer to Figure 2-13; the selected object will be displayed in transparent lines and the **Select objects** prompt will be displayed again. You can either continue selecting the objects or press ENTER to terminate the object selection process and erase the selected objects.

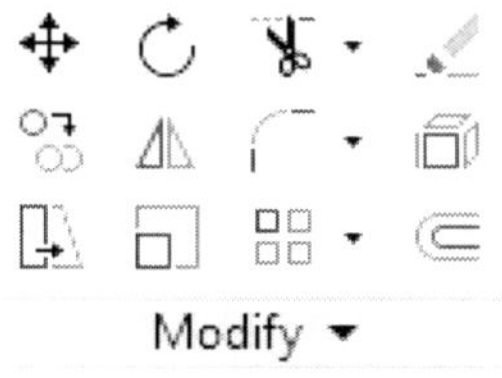

***Figure 2-11** The **Erase** tool in the **Modify** panel*

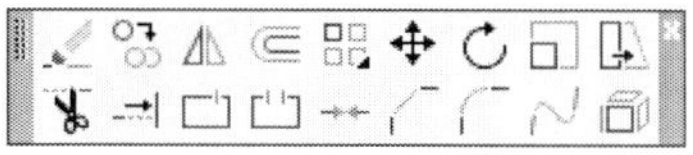

***Figure 2-12** The **Erase** tool in the **Modify** toolbar*

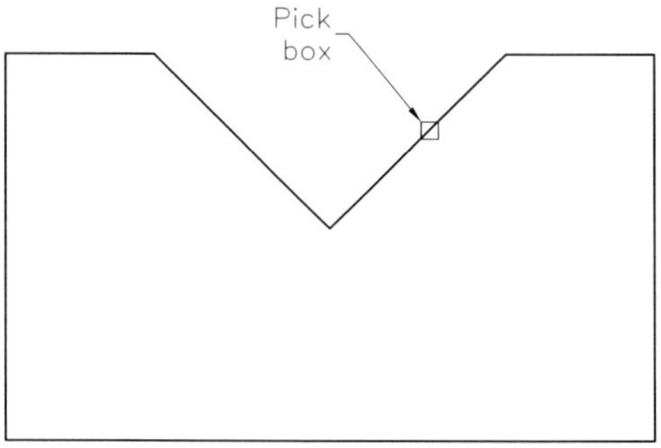

***Figure 2-13** Selecting the object by positioning the pick box at the top of the object*

DRAWING A CIRCLE

Ribbon: Home > Draw > Circle
Toolbar: Draw > Circle **Menu Bar:** Draw > Circle
Tool Palettes: Draw > Circle **Command:** CIRCLE/C

In AutoCAD, you can draw a circle by using six different tools. All these tools are grouped together in the **Draw** panel of the **Ribbon**. To view these tools, choose the down arrow below the **Circle** tool in the **Draw** panel, as shown in Figure 2-14; all tools will be listed in the drop-down. Note that the name of the tool chosen last will be displayed in the **Draw** panel. You can also invoke the **Circle** tool from the **Draw** tab in the **Tool Palettes** or by entering **C** in the command prompt.

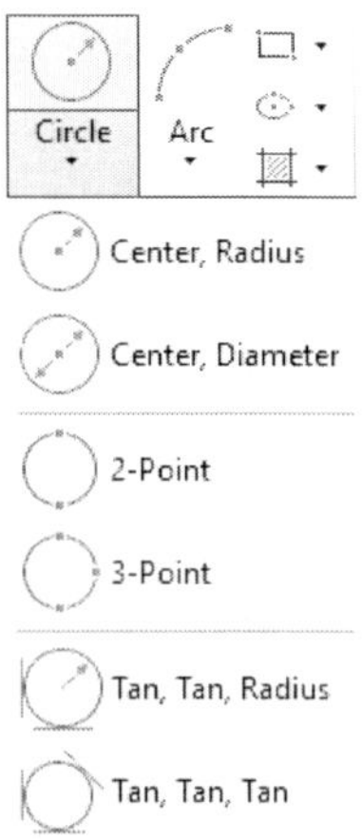

***Figure 2-14** Tools in the **Circle** drop-down*

EXERCISES

Exercise 1 *Line and Circle*

Draw the profile shown in Figure 2-15 using various options of the **Line** and **Circle** tools. Use the absolute, relative rectangular, or relative polar coordinates for drawing the triangle. The vertices of the triangle will be used as the centers of the circles. The circles can be drawn by using the **Center, Radius**, or **Center, Diameter**, or **Tan, Tan, Tan** tools.

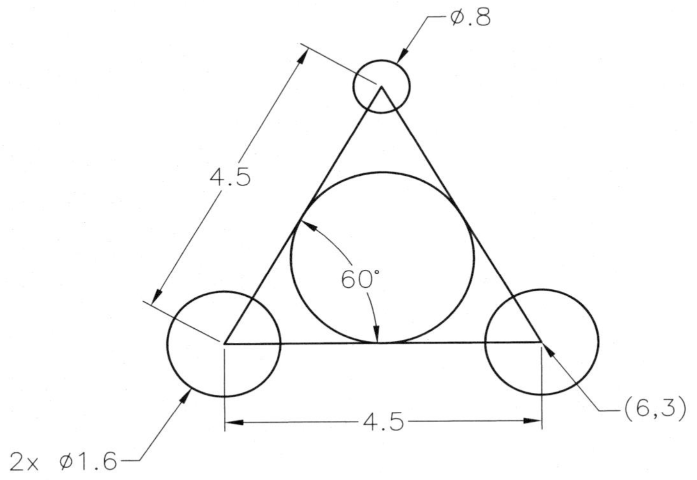

Figure 2-15 *Drawing for Exercise 1*

Exercise 2 *Relative Rectangular & Polar Coordinates*

Draw the profile shown in Figure 2-16 by using the relative rectangular and relative polar coordinates of the points given in the following table. The distance between the dotted lines is 1 unit. Save this drawing with the name *C02_Exer02.dwg*.

Point	Coordinates	Point	Coordinates
1	3.0, 1.0	9	______
2	______	10	______
3	______	11	______
4	______	12	______
5	______	13	______
6	______	14	______
7	______	15	______
8	______	16	______

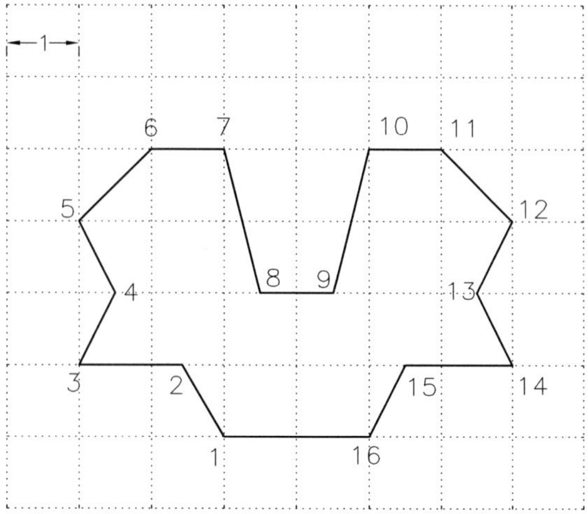

Figure 2-16 *Drawing for Exercise 2*

Chapter 3

Getting Started with Advanced Sketching

Learning Objectives

After completing this chapter, you will be able to:

- *Draw arcs using various options*
- *Draw rectangles, ellipses, elliptical arcs, and polygons*
- *Draw polylines and donuts*

DRAWING ARCS

Ribbon: Home > Draw > Arc drop-down **Toolbar:** Draw > Arc
Menu Bar: Draw > Arc **Command:** ARC/A

An arc is defined as a segment of a circle. In AutoCAD, an arc is drawn by using the tools available in the **Arc** drop-down of the **Draw** panel in the **Ribbon**, refer to Figure 3-1. You can choose a tool on the basis of known parameters. Remember that the tool that was used last to create an arc will be displayed in the **Draw** panel. The methods to draw an arc are discussed next.

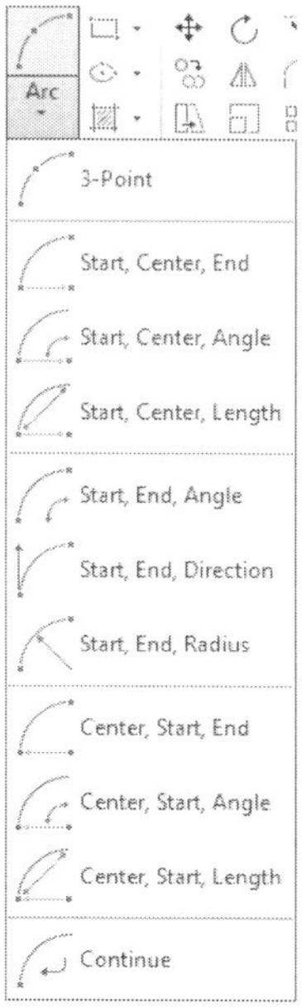

***Figure 3-1** The tools in the **Arc** drop-down*

Drawing an Arc by Specifying Three Points

To draw an arc by specifying the start point, endpoint, and another point on its periphery, choose the **3-Point** tool from the **Arc** drop-down in the **Draw** panel, refer to Figure 3-1. On doing so, you will be prompted to specify the start point. Specify the first point or specify coordinates for the first point. Then, specify the second point and endpoint of the arc.

Drawing an Arc by Specifying its Start Point, Center Point, and Endpoint

If you know the start point, endpoint, and center point of an arc, choose the **Start, Center, End** tool from the **Arc** drop-down in the **Draw** panel and then specify the start, center, and end points in succession; the arc will be drawn. The radius of the arc is determined by the distance between the center point and the start point. Therefore, the endpoint is used to calculate the angle at which the arc ends.

Drawing an Arc by Specifying its Start Point, Center Point, and Included Angle

Included angle is the angle between the start and end points of an arc about the specified center. If you know the location of the start point, center point, and included angle of an arc, choose the **Start, Center, Angle** tool from the **Arc** drop-down in the **Draw** panel and specify the start point, center point, and included angle; the arc will be drawn in a counterclockwise direction with respect to the specified center and start point.

Drawing an Arc by Specifying the Start Point, Center Point, and Chord Length

A chord is defined as a straight line connecting the start point and endpoint of an arc. To draw an arc by specifying its chord length, choose the **Start, Center, Length** tool from the **Arc** drop-down in the **Draw** panel and specify the start point, center point, and length of the chord in succession. On specifying the chord length, AutoCAD will calculate the included angle and an arc will be drawn in the counterclockwise direction from the start point.

Drawing an Arc by Specifying its Start Point, Endpoint, and Included Angle

To draw an arc by specifying its start point, endpoint, and the included angle, choose the **Start, End, Angle** tool from the **Arc** drop-down in the **Draw** panel and specify the start point, endpoint, and the included angle in succession; the arc will be drawn.

Drawing an Arc by Specifying its Start Point, Endpoint, and Direction

This option is used to draw a major or minor arc, whose size and position are determined by the distance between the start point and endpoint and the direction specified. You can specify the direction by selecting a point on a line that is tangent to the start point or by entering an angle between the start point of the arc and the end point of the tangent line.

To draw an arc by specifying its direction, choose the **Start, End, Direction** tool from the **Arc** drop-down in the **Draw** panel and specify the start and end points in succession; you will be prompted to specify the direction. Specify a point on the line that is tangent to the start point or enter an angle between the start point of the arc and the end point of the tangent line; an arc will be drawn.

Drawing an Arc by Specifying its Start Point, Endpoint, and Radius

If you know the location of the start point, endpoint, and radius of an arc, choose the **Start, End, Radius** tool from the **Arc** drop-down in the **Draw** panel and specify the start and end points; you will be prompted to specify the radius. Enter the radius value; the arc will be drawn.

Drawing an Arc by Specifying its Center Point, Start Point, and Endpoint

The **Center, Start, End** tool is the modification of the **Start, Center, End** tool. Use this tool, whenever it is easier to start drawing an arc by establishing the center first. Here, the arc is always drawn in a counterclockwise direction from the start point to the endpoint, around the specified center.

Drawing an Arc by Specifying its Center Point, Start Point, and Angle

You can use the **Center, Start, Angle** tool if you need to draw an arc by specifying the center first.

Drawing an Arc by Specifying the Center Point, Start Point, and Chord Length

The **Center, Start, Length** tool is used whenever it is easier to draw an arc by establishing the center first.

Drawing an Arc by Using the Continue Tool

To create an arc that is tangent to a previously drawn arc or line, choose the **Continue** tool from the **Arc** drop-down in the **Draw** panel; the start point and the direction of the arc will be taken from the endpoint and the ending direction of the previous line or arc. Next, specify the endpoint to draw an arc.

DRAWING RECTANGLES

Ribbon: Home > Draw > Rectangle **Toolbar:** Draw > Rectangle
Tool Palettes: Draw > Rectangle **Command:** RECTANG/REC

A rectangle is drawn by choosing the **Rectangle** tool from the **Draw** panel of the **Home** tab. In AutoCAD, you can draw rectangles by specifying two opposite corners of the rectangle by specifying the area and the size of one of the sides or by specifying the dimensions of the rectangle. All these methods of drawing rectangles are discussed next.

Drawing Rectangles by Specifying Two Opposite Corners

On invoking the **Rectangle** tool, you will be prompted to specify the first corner of the rectangle. Enter the coordinates of the first corner or specify the first corner by using the mouse; you will be prompted to specify the other corner. The first corner can be any one of the four corners. Specify the diagonally opposite corner by entering the coordinates or by using the left mouse button.

Drawing Rectangles by Specifying the Area and One Side

To draw a rectangle by specifying its area and the length of one of the sides, first specify the start point. Next, invoke the shortcut menu by right-clicking and then choose the **Area** option. Next, specify the parameters; the rectangle is drawn.

Drawing Rectangles by Specifying their Dimensions

You can also draw a rectangle by specifying its dimensions. This can be done by choosing the **Dimensions** option from the shortcut menu at the **Specify other corner point or [Area/Dimensions/Rotation]** prompt and entering the length and width of the rectangle.

Drawing Rectangle at an Angle

You can also draw a rectangle at an angle. This can be done by choosing the **Rotate** option from the shortcut menu, at the **Specify other corner point or [Area/Dimensions/Rotation]** prompt and entering the rotation angle. After entering the rotation angle, you can continue sizing the rectangle using any one of the above discussed methods.

DRAWING ELLIPSES

Ribbon: Home > Draw > Ellipse drop-down **Toolbar:** Draw > Ellipse
Tool Palettes: Draw > Ellipse **Command:** ELLIPSE/EL

If you cut a cone by a cutting plane at an angle and view the cone perpendicular to the cutting plane, the shape created is called an ellipse. An ellipse can be created by using different tools available in the **Ellipse** drop-down of the **Draw** panel.

Once you invoke the **Ellipse** tool, the **Specify axis endpoint of ellipse or [Arc/Center]** or **Specify axis endpoint of ellipse or [Arc/Center/Isocircle]** (if isometric snap is ON) prompt will be displayed. The response to this prompt depends on the option you choose. The various options are explained next.

Note

*By default, the **Isocircle** option is not available for the **Ellipse** tool. To display this option, you have to select the **Isometric snap** radio button in the **Snap and Grid** tab of the **Drafting Settings** dialog box or choose the **Isometric Drafting** button in the Status Bar.*

Drawing Ellipse Using the Center Option

To draw an ellipse by specifying its center point, endpoint of one of its axes, and endpoint of other axis, choose the **Center** tool from the **Ellipse** drop-down in the **Draw** panel; you will be prompted to specify the center of the ellipse. The center of an ellipse is defined as the point of intersection of the major and minor axes. Specify the center point or enter coordinates; you will be prompted to specify the endpoint. Specify the endpoint of the major or minor axis; you will be prompted to specify the distance of the other axis. Specify the distance; the ellipse will be drawn.

Drawing an Ellipse by Specifying its Axis and Endpoint

To draw an ellipse by specifying one of its axes and the endpoint of the other axis, choose the **Axis, End** tool from the **Ellipse** drop-down of the **Draw** panel; you will be prompted to specify the axis end point. Specify the first endpoint of one axis of the ellipse; you will be prompted to specify the other endpoint of the axis. Specify the other endpoint of the axis. Now, you can specify the distance to other axis from the center or specify the rotation angle around the specified axis.

Drawing Elliptical Arcs

Ribbon: Home > Draw > Ellipse drop-down > Elliptical Arc	**Toolbar:** Draw > Ellipse Arc
Tool Palettes: Draw > Ellipse Arc	**Command:** ELLIPSE > Arc

In AutoCAD, you can draw an elliptical arc by choosing the **Elliptical Arc** tool from the **Ellipse** drop-down of the **Draw** panel. On choosing this tool, you can specify the endpoints of one of the axes, the distance to other axis from the center and any one of the following information:

1. Start and End angles of the arc.
2. Start and Included angles of the arc.
3. Start and End parameters.

DRAWING REGULAR POLYGONS

Ribbon: Home > Draw > Rectangle drop-down > Polygon	**Toolbar:** Draw > Polygon
Tool Palettes: Draw > Polygon	**Command:** POLYGON/POL

A regular polygon is a closed geometric entity with equal sides. The number of sides of a polygon vary from 3 to 1024. For example, a triangle is a three-sided polygon and a pentagon is a five-sided polygon. To draw a regular 2D polygon, choose the **Polygon** tool from the **Rectangle** drop-down of the **Draw** panel; you will be prompted to specify the number of sides. Type the number of sides and press ENTER. Now, you can draw the polygon by specifying the length of an edge or by specifying the center of the polygon.

DRAWING POLYLINES

Ribbon: Home > Draw > Polyline **Toolbar:** Draw > Polyline
Tool Palettes: Draw > Polyline **Command:** PLINE/PL

To draw a polyline, choose the **Polyline** tool from the **Draw** panel. The **Polyline** tool functions fundamentally like the **Line** tool except that some additional options are provided and all segments of the polyline form a single object. After invoking the **Polyline** tool, the following prompt is displayed.

Specify start point: *Specify the starting point or enter its coordinates.*
Current line-width is 0.0000
Specify next point or [Arc/Halfwidth/Length/Undo/Width]:

Note that the message **Current line-width is 0.0000** is displayed automatically indicating that the polyline drawn will have 0.0000 width.

When you are prompted to specify the next point, you can continue specifying the next point and draw a polyline, or depending on your requirements, the other options can be invoked.

DRAWING DONUTS

Ribbon: Home > Draw > Donut **Command:** DONUT/DOUGHNUT/DO

In AutoCAD, the **Donut** tool is used to draw an object that looks like a filled circular ring called donut. AutoCAD's donuts are made of two semicircular polyarcs with a certain width. Therefore, the **Donut** tool allows you to draw a thick circle. The donuts can have any inside and outside diameters. If **FILLMODE** is off, a donut will look like a circle (if the inside diameter is zero) or a concentric circle (if the inside diameter is not zero). On invoking the **Donut** tool, you will be prompted to specify the diameters. After specifying the two diameters, the donut gets attached to the crosshairs. Specify a point for the center of the donut in the drawing area to place the donut. In this way, you can place as many donuts as required without exiting the tool. Press ENTER to exit the tool.

The default values for the inside and outside diameters of donuts are saved in the **DONUTID** and **DONUTOD** system variables. A solid-filled circle is drawn by specifying the inside diameter as zero, if **FILLMODE** is on.

Tutorial 1 — Donut

You will draw an unfilled donut shown in Figure 3-2 with an inside diameter of 0.75 unit, an outside diameter of 2.0 units, and centered at (2,2). You will also draw a filled donut and a solid-filled donut with the given specifications.

The following is the prompt sequence to draw an unfilled donut shown in Figure 3-2.

Command: **FILLMODE** [Enter]
New value for FILLMODE <1>: **0** [Enter]
Command: *Choose the **Donut** tool from the **Draw** panel*
Specify inside diameter of donut<0.5000>: **0.75** [Enter]
Specify outside diameter of donut <1.000>: **2** [Enter]
Specify center of donut or <exit>: **2,2** [Enter]
Specify center of donut or <exit>: [Enter]

The prompt sequence for drawing a filled donut with an inside diameter of 0.5 unit, outside diameter of 2.0 units, which is centered at a specified point is given below:

Command: **FILLMODE** [Enter]
Enter new value for FILLMODE <0>: **1** [Enter]
Command: *Choose the **Donut** tool from the **Draw** panel*
Specify inside diameter of donut<0.5000>: **0.5** [Enter]
Specify outside diameter of donut <1.000>: **2** [Enter]
Specify center of donut or <exit>: *Specify a point*
Specify center of donut or <exit>: [Enter]. *Refer to Figure 3-3*

To draw a solid-filled donut with an outside diameter of 2.0 units, use the following prompt sequence:

*Choose the **Donut** tool from the **Draw** panel*
Specify inside diameter of donut <0.50>: **0** [Enter]
Specify outside diameter of donut <1.0>: **2** [Enter]
Specify center of donut or <exit>: *Specify a point*
Specify center of donut or <exit>: [Enter]. *Refer to Figure 3-4*

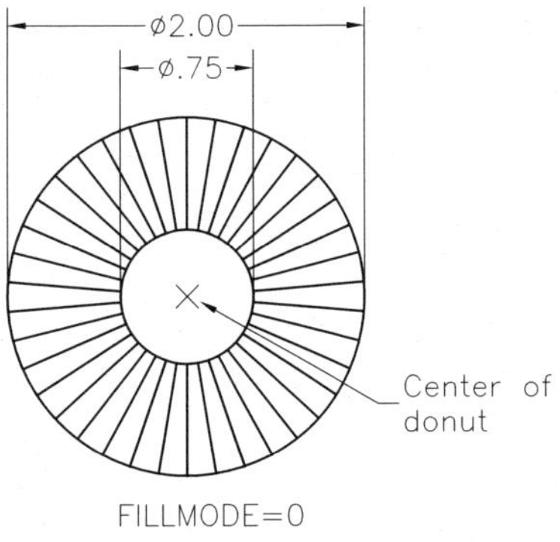

Figure 3-2 *Unfilled donut*

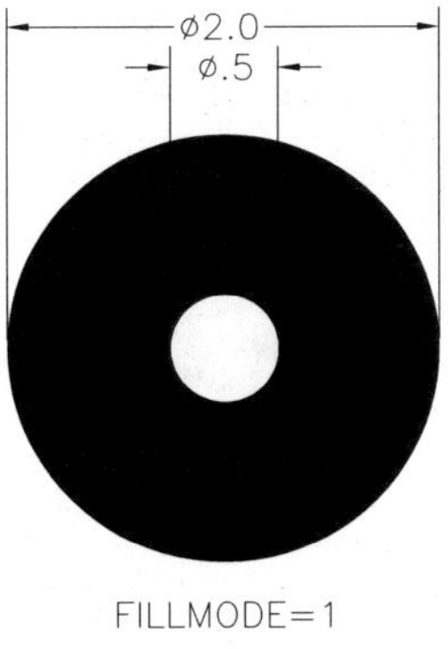

Figure 3-3 *Filled donut*

Figure 3-4 *Solid-filled donut*

EXERCISES

Exercise 1 — Start, Center, Angle

a. Draw an arc whose start point is at 6,3, center point is at 3,3, and the included angle is 240 degrees.

b. Draw the profile shown in Figure 3-5. The distance between the dotted lines is 1.0 unit. Create the arcs by using different arc command options as indicated in the figure.

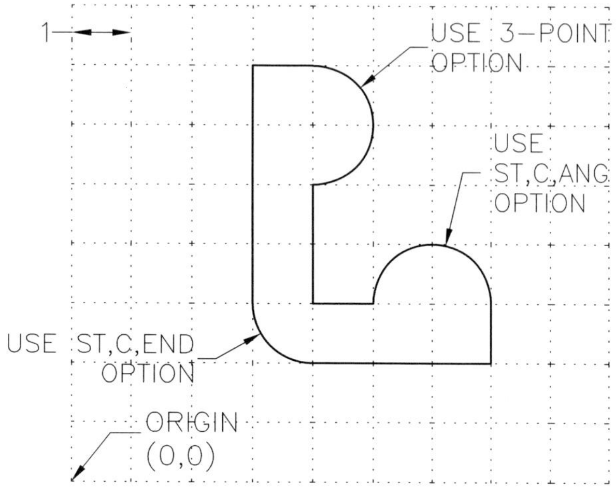

Figure 3-5 *Drawing for Exercise 1*

Exercise 2 — Elliptical Arc

a. Construct an ellipse with center at (2,3), axis endpoint at (4,6), and the other axis endpoint at a distance of 0.75 unit from the midpoint of the first axis.

b. Draw the profile shown in Figure 3-6. The distance between the dotted lines is 1.0 unit.

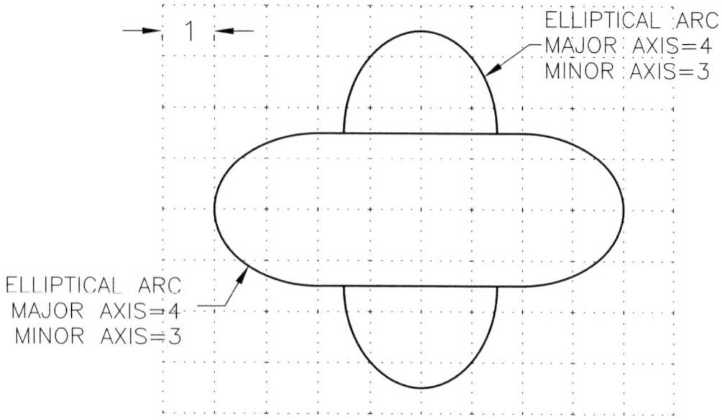

Figure 3-6 *Drawing for Exercise 2*

Chapter 4

Working with Drawing Aids

Learning Objectives

After completing this chapter, you will be able to:

- *Set up layers, and assign colors and linetypes to them*

INTRODUCTION

In this chapter, you will learn about the drawing setup and the factors that affect the quality and accuracy of a drawing. This chapter contains a detailed description of the procedure of setting up a layer. You will also learn about some other drawing aids, such as Grid, Snap, and Ortho. These aids will help you create drawings accurately and quickly.

UNDERSTANDING THE CONCEPT AND USE OF LAYERS

The concept of layers can be best explained by using the concept of overlays in manual drafting. In manual drafting, different details of a drawing can be drawn on different sheets of paper or overlays. Each overlay is perfectly aligned with the others. Once all of them are placed on top of each other, you can reproduce the entire drawing. In AutoCAD, instead of using overlays, you can use layers. Each layer is assigned a name. You can also assign a color and a line type to a layer.

Advantages of Using Layers

1. Each layer can be assigned a different color. Assigning a particular color to a group of objects is very important for plotting. For example, if all object lines are red, then at the time of plotting, you can assign the red color to a slot (pen) that has the desired tip width. Similarly, if the dimensions are green, you can assign the green color to another slot (pen) that has a thin tip. By assigning different colors to different layers, you can control the width of lines while plotting the drawing. You can also make a layer plotable or non-plotable.
2. Layers are also useful for performing some editing operations. For example, to erase all dimensions in a drawing, you can freeze or lock all layers except the dimension layer, then select all objects by using the Window Crossing option, and erase all dimensions.
3. You can turn off a layer or freeze a layer that you do not want to be displayed or plotted.
4. You can lock a layer to prevent the user from accidentally editing the objects in it.
5. Colors also help you distinguish different groups of objects. For example, in architectural drafting, the plans for foundation, floors, plumbing, electrical work, and heating systems may be made in different layers. In electronic drafting and in PCB (printed circuit board), the design of each level of a multilevel circuit board can be drawn on a separate layer. Similarly, in mechanical engineering, the main components of an assembly can be made in one layer, other components such as nuts, bolts, keys, and washers can be made in another layer, and the annotations such as datum symbols and identifiers, texture symbols, Balloons, and Bill of Materials can be made in yet another layer.

WORKING WITH LAYERS

Enhanced

Ribbon: Home > Layers > Layer Properties **Command:** LAYER/LA
Toolbar: Layers > Layer Properties Manager

Layer Properties

You can freeze, thaw, lock, unlock layers using the **Layers** panel in the **Ribbon**. The **Layers** panel is shown in Figure 4-1. However, to add new layers, to delete the existing layers, or to assign colors and linetypes to layers, you need to invoke the **LAYER PROPERTIES MANAGER.**

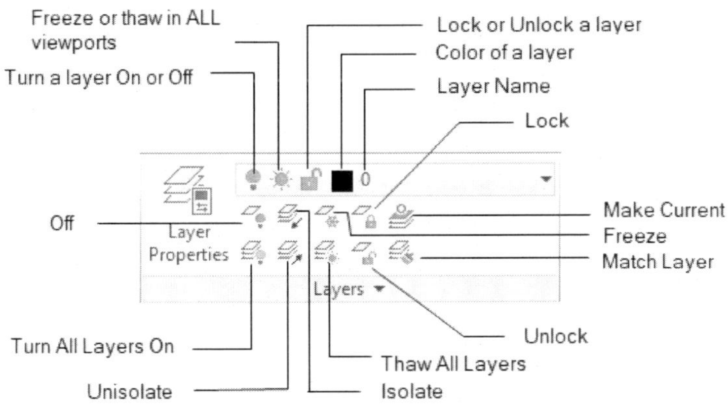

*Figure 4-1 The **Layers** panel*

To invoke the **LAYER PROPERTIES MANAGER**, choose the **Layer Properties** button from the **Layers** panel of the **Home** tab, or invoke it from the **Layers** toolbar, as shown in Figure 4-2.

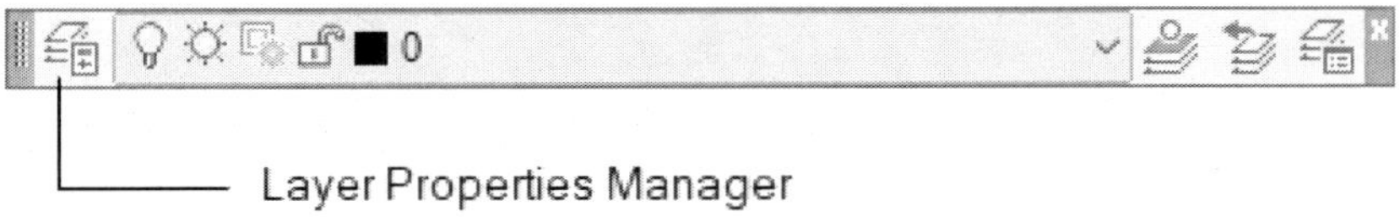

*Figure 4-2 The **Layers** toolbar*

On invoking the **LAYER PROPERTIES MANAGER**, a default layer with the name 0 is displayed. It is the current layer and any object you draw is created in it. The current layer can be recognized by the green colored tick mark in the **Status** column. There are certain features such as color, linetype, and lineweight that are associated with each layer. The **0** layer has the default color as white, linetype as continuous, and lineweight as default.

Creating New Layers

To create new layers, choose the **New Layer** button in the **LAYER PROPERTIES MANAGER**; a new layer, named Layer1, with the same properties as that of the current layer will be created and listed, as shown in Figure 4-3.

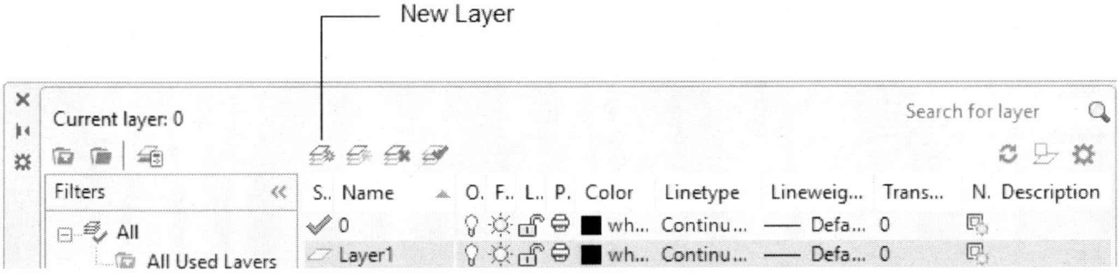

***Figure 4-3** Partial view of the **LAYER PROPERTIES MANAGER** with new layers created*

Naming a New Layer

New layers are created with the name Layer 1, Layer 2, and so on. To change or edit the name of a layer, select it, then click once in the field corresponding to the **Name** column, and enter a new name.

Making a Layer Current

To draw an object in a particular layer, you need to make it the current layer. There are different methods to make a layer current and these are listed below.

1. Double-click on the name of a layer in the list box in the **LAYER PROPERTIES MANAGER**; the selected layer is made current.
2. Select the name of the layer in the **LAYER PROPERTIES MANAGER** and then choose the **Set Current** button in the **LAYER PROPERTIES MANAGER**.
3. Right-click on a layer in the **Layer** list box in the **LAYER PROPERTIES MANAGER** and choose the **Set current** option from the shortcut menu displayed, refer to Figure 4-4.

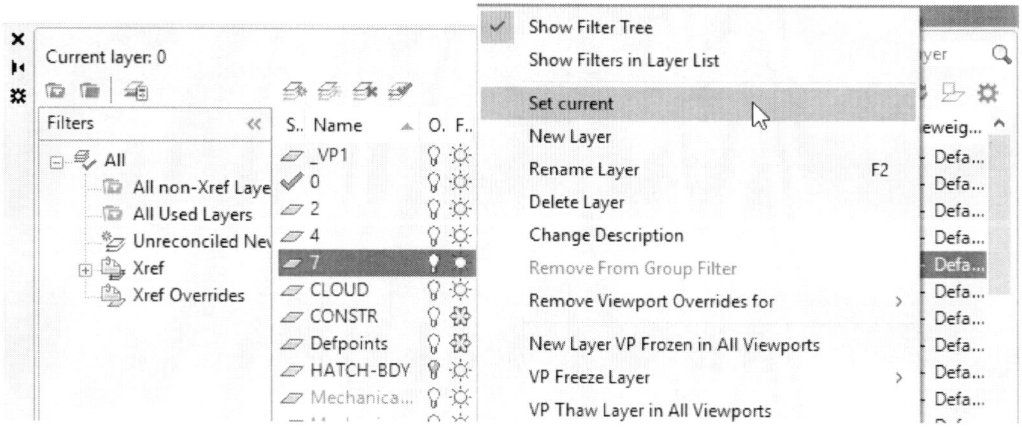

Figure 4-4 Partial view of the ***LAYER PROPERTIES MANAGER*** *with the shortcut menu*

Controlling the Display of Layers

You can control the display of layers by using the display options from the **LAYER PROPERTIES MANAGER**. These options are discussed next.

Turn a layer On or Off

Choose the **Turn a layer On or Off** toggle icon (bulb) to turn a layer on or off. You can also turn the layer on or off by clicking on the **Turn a layer On or Off** toggle icon located in the **Layer** drop-down list from the **Layers** panel or from the **Quick Access Toolbar,** as shown in Figure 4-5.

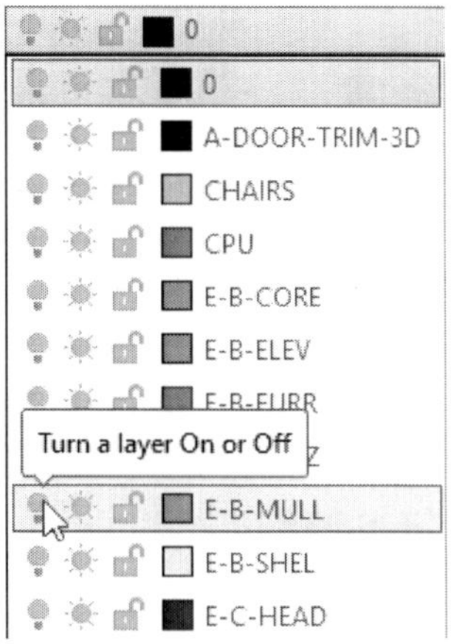

Figure 4-5 *Turning off the* ***E-B-MULL*** *layer*

Freeze or thaw in ALL Viewports

Sometimes, in an architectural drawing, you may not need the door tag, window tag, or surveyor data to be displayed, so that they are not changed. These types of information or entities can be placed in a particular layer and that layer can be frozen. You cannot edit the entities in the frozen layer. Also, frozen layers are invisible and cannot be plotted. To freeze a layer, choose the **Freeze or thaw in ALL viewports** toggle icon (sun/snowflakes) in the **LAYER PROPERTIES MANAGER**. Note that the current layer cannot be frozen.

New VP Freeze

On choosing the **Layout** tab, you will observe that a default viewport is displayed in the drawing area. You can also create new viewports anytime during the design. If you want to freeze some layers in all the subsequent new viewports, select the layers, and then choose the **New VP Freeze** toggle icon. The selected layers will be frozen in all the subsequently created viewports, without affecting the existing viewports.

VP Freeze

This icon will be available only if you invoke the **LAYER PROPERTIES MANAGER** in the **Layout** tab, refer to Figure 4-6. The **Freeze or thaw in current viewport** toggle icon will also be available in the **Layer** drop-down list in the **Layers** panel of the **Ribbon** or in the **Layers** toolbar. If there are multiple viewports, make a viewport as the current viewport by double-clicking in it and freeze or thaw the selected layer in it by choosing the **VP Freeze** icon in the **LAYER PROPERTIES MANAGER**. However, a layer that is frozen in the model space cannot be thawed in the current viewport.

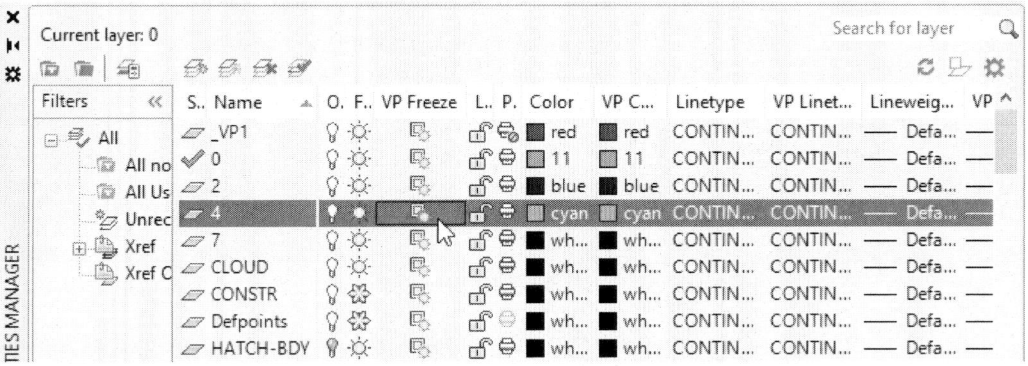

*Figure 4-6 Partial view of the **LAYER PROPERTIES MANAGER** with the **VP Freeze** icon*

Lock or Unlock a Layer

While working on a drawing, if you want to avoid editing some objects on a particular layer but need to have them visible, use the Lock/Unlock toggle icon to lock the layer. The **Unlock** option overrules the **Lock** option and allows you to edit objects on the layers that were previously locked.

Make a Layer Plotable or Non plotable

By default, you can plot all layers, except the layers that are turned off or frozen. If you do not want to plot a layer that is not turned off or frozen, select the layer and choose the **Plot** icon (printer). This is a toggle icon, therefore you can choose this icon again to plot the layer.

Assigning Linetype to a Layer

By default, continuous linetype is assigned to a layer, if no layer is selected while creating a new layer. Otherwise, new layer takes the properties of the selected layer. To assign a new linetype to a layer, click on the name of the linetype under the **Linetype** column of that layer in the **LAYER PROPERTIES MANAGER**; the **Select Linetype** dialog box with the linetypes loaded on your computer will be displayed. Select the new linetype and then choose the **OK** button; the selected linetype will be assigned to the layer. If you are opening the **Select Linetype** dialog box for the first time, only the **Continuous** linetype will be displayed, as shown in Figure 4-7.

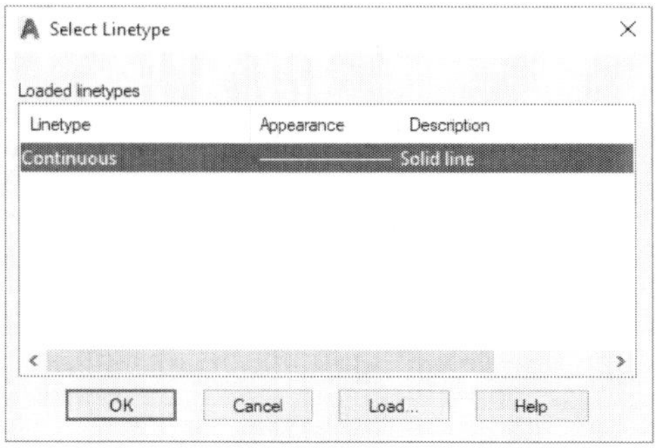

Figure 4-7 *The* ***Select Linetype*** *dialog box*

You need to load linetypes and then assign them to layers. To load linetypes, choose the **Load** button in the **Select Linetype** dialog box; the **Load or Reload Linetypes** dialog box will be displayed, refer to Figure 4-8. This dialog box displays all linetypes in the *acad.lin* or *acadiso.lin* file. In this dialog box, you can select a single linetype, or a number of linetypes by pressing and holding the SHIFT or CTRL key and then selecting the linetypes. After selecting the linetypes, choose the **OK** button; the selected linetypes are loaded in the **Select Linetype** dialog box. Now, select the desired linetype and choose the **OK** button; the selected linetype is assigned to the selected layer.

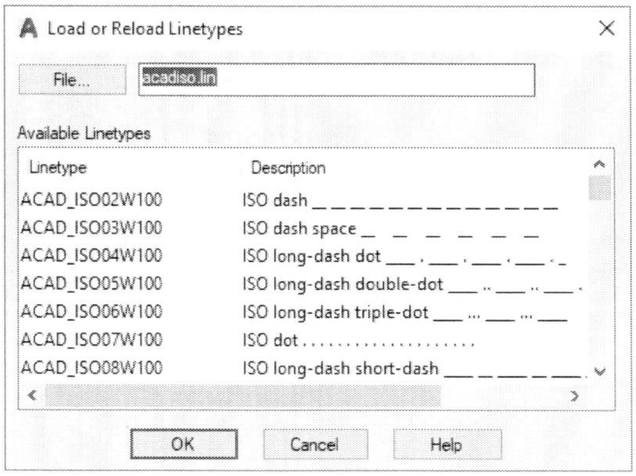

Figure 4-8 *The* ***Load or Reload Linetypes*** *dialog box*

Assigning Transparency to a Layer

By default, no transparency is assigned to a layer, if **Layer 0** is selected while creating that layer. Otherwise, new layer takes the transparency of the selected layer. To assign transparency to a layer, click on the **Transparency** field of that layer; the **Layer Transparency** dialog box will be displayed. Select the required transparency level from the **Transparency value (0-90)** drop-down list and choose **OK**. You can also enter a value in the edit box. Now, if you place an object on this layer, the object will have a faded color according to the specified transparency value.

Assigning Color to a Layer

To assign a color to a layer, select the color swatch in that layer in the **LAYER PROPERTIES MANAGER**; the **Select Color** dialog box will be displayed. Select the desired color and then choose the **OK** button; the selected color will be assigned to the layer.

Assigning Lineweight to a Layer

Lineweight is used to give thickness to objects in a layer. For example, if you create a sectional plan, you can assign a layer with a larger value of lineweight to create the objects through which the section is made. Another layer with a lesser lineweight can be used to show the objects through which the section does not pass. This thickness is displayed on the screen if the display of the lineweight is on. The lineweight assigned to an object can also be plotted. To assign a lineweight to a layer, select the layer and then click on the lineweight associated with it; the **Lineweight** dialog box will be displayed. Select a lineweight and then choose the **OK** button from this dialog box to return to the **LAYER PROPERTIES MANAGER**.

Deleting Layers

To delete a layer, select it and then choose the **Delete Layer** button from the **LAYER PROPERTIES MANAGER**. Note that you cannot delete following layers: 0, Defpoints (created while dimensioning), Ashade (created while rendering), current layer, a layer that contains an object, and an Xref-dependent layer.

Tutorial 1 — *Layers*

Set up three layers with the following linetypes and colors. Then, create the drawing shown in Figure 4-9 (without dimensions).

Layer Name	Color	Linetype	Lineweight
Obj	Red	Continuous	0.012”
Hid	Yellow	Hidden	0.008”
Cen	Green	Center	0.006”

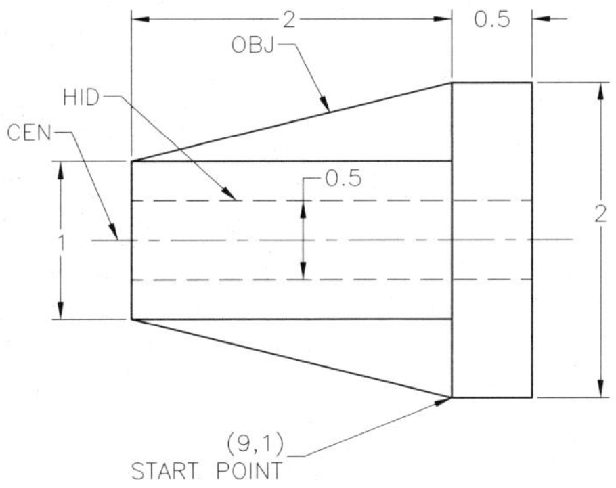

Figure 4-9 *Drawing for Tutorial 1*

In this tutorial, assume that the limits and units are already set. Before drawing the lines, you need to create layers and assign colors, linetypes, and lineweights to them. Also, depending on the objects that you want to draw, you need to set a layer as the current layer. In this tutorial, you will create layers by using the **LAYER PROPERTIES MANAGER**. You will use the **Layer** drop-down list in the **Layers** panel to set a layer as the current layer and then draw the figure.

1. As the lineweights specified are given in inches, first you need to change their units, if they are in millimeters. To do so, invoke the **Lineweight Settings** dialog box by right-clicking on the **Show/Hide Lineweight** button on the Status Bar and then choose the **Lineweight Settings** option from the shortcut menu. Select the **Inches (in)** radio button in the **Units for Listing** area of the dialog box and then choose the **OK** button.

2. Choose the **Layer Properties** button from the **Layers** panel in the **Home** tab of the **Ribbon** to display the **LAYER PROPERTIES MANAGER**. The layer **0** with default properties is displayed in the list box.

3. Choose the **New Layer** button; a new layer (Layer1) with the default properties is displayed in the list box. Change the default name, **Layer1** to **Obj**.

4. Left-click on the **Color** field of this layer; the **Select Color** dialog box is displayed. Select the **Red** color and then choose the **OK** button; red color is assigned to the **Obj** layer.

5. Left-click on the **Lineweight** field of the **Obj** layer; the **Lineweight** dialog box is displayed. Select **0.012"** and then choose the **OK** button; the selected lineweight is assigned to the **Obj** layer.

6. Again, choose the **New Layer** button; a new layer (Layer1) with properties similar to that of the **Obj** layer is created. Change the default name to **Hid**.

7. Left-click on the **Color** field of this layer; the **Select Color** dialog box is displayed. Select the **Yellow** color and then choose the **OK** button.

8. Left-click on the **Linetype** field of the **Hid** layer; the **Select Linetype** dialog box is displayed. If the **HIDDEN** linetype is not displayed in the dialog box, choose the **Load** button; the **Load or Reload Linetypes** dialog box is displayed. Select **HIDDEN** from the list and choose the **OK** button from the **Load or Reload Linetypes** dialog box. Next, select **HIDDEN** from the **Select Linetype** dialog box and then choose the **OK** button.

9. Left-click on the **Lineweight** field of the **Hid** layer; the **Lineweight** dialog box is displayed. Select **0.008"** and then choose the **OK** button; the selected lineweight is assigned to the **Hid** layer.

10. Similarly, create the new layer **Cen** and assign the color **Green**, linetype **CENTER**, and lineweight **0.006"** to it.

11. Select the **Obj** layer and then choose the **Set Current** button to make the **Obj** layer current, refer to Figure 4-10. Choose the **Close** button to exit the **LAYER PROPERTIES MANAGER**.

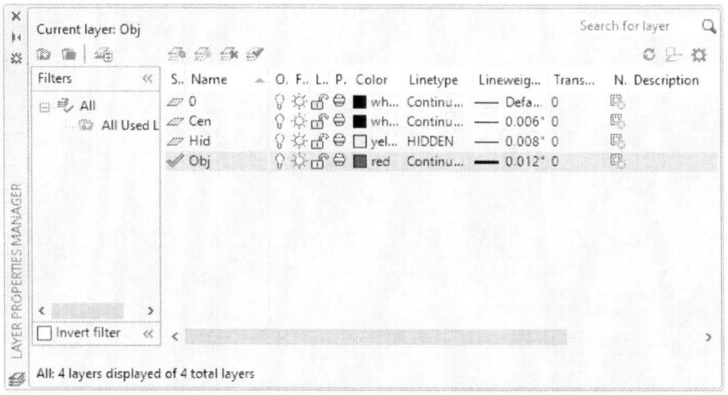

Figure 4-10 *Layers created for Tutorial 1*

12. Choose the **Show/Hide Lineweight** button from the Status Bar to turn on the display of the lineweights of the lines to be drawn.

13. Choose the **Dynamic Input** button from the Status Bar to turn it on.

14. Choose the **Line** tool from the **Draw** panel and draw solid lines in the drawing. Make sure that the start point of the line is at 9,1. You will notice that a continuous line is drawn in red color. This is because the **Obj** layer is the current layer and red color is assigned to it.

15. Click the down arrow in the **Layer** drop-down list available in the **Layers** panel to display the list of layers and then select the **Hid** layer from the list to make it current, as shown in Figure 4-11.

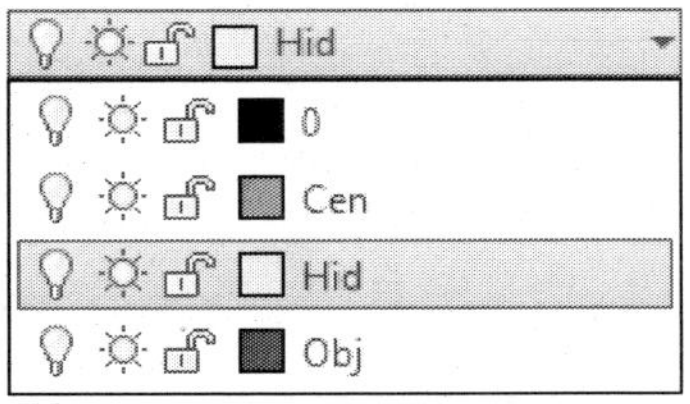

Figure 4-11 *Selecting the **Hid** layer from the **Layers** panel to set it as the current layer*

16. Draw two hidden lines; the lines are displayed in yellow color with hidden linetype.

17. Draw the center line; the centerline is displayed in yellow color with hidden linetype. This is because the **Hid** layer is the current layer.

18. Now, select the centerline and then select the **Cen** layer from the **Layer** drop-down list available in the **Layers** panel; the color and linetype of the centerline are changed.

EXERCISES

Exercise 1 — Layers

Set up layers with the following linetypes and colors. Then make the drawing (without dimensions), as shown in Figure 4-12. The distance between the dotted lines is 1.0 unit.

Layer Name	Color	Linetype
Object	Red	Continuous
Hidden	Yellow	Hidden
Center	Green	Center
Dimension	Blue	Continuous

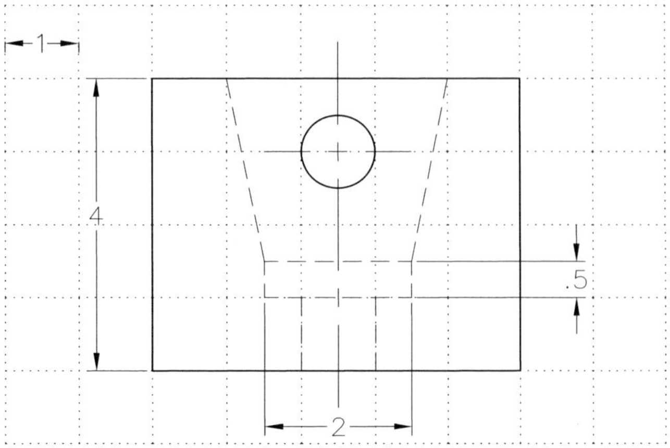

Figure 4-12 *Drawing for Exercise 1*

Exercise 2 — Line type and Object Color

Set up layers with the following linetypes and colors. Then make the drawing, as shown in Figure 4-13. The distance between the dotted lines is 1.0 unit.

Layer name	Color	Linetype
Object	Red	Continuous
Hidden	Yellow	Hidden
Center	Green	Center

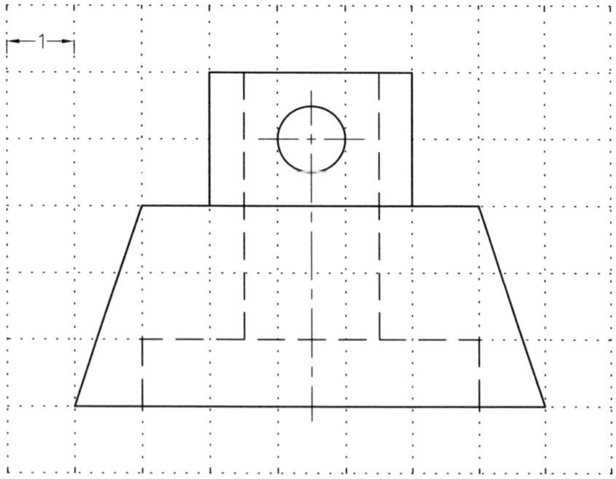

Figure 4-13 *Drawing for Exercise 2*

Chapter 5

Editing Sketched Objects-I

Learning Objectives

After completing this chapter, you will be able to:

- *Create selection sets using various object selection options*
- *Use the Move and Copy tools*
- *Copy objects using the Array tool*
- *Understand various editing tools*

CREATING A SELECTION SET

For most of the tools, the default object selection method is to use the pick box (cursor) and select one entity at a time. You can also click in blank area and select objects by using the **Window** option or the **Crossing** option. In Chapter 2, these two options were discussed.

EDITING SKETCHES

To use AutoCAD efficiently, you need to know the editing tools and how to use them. In this section, you will learn about the editing tools. These tools can be invoked from the **Ribbon**, or toolbar, or by entering them at the command prompt.

MOVING SKETCHED OBJECTS

Ribbon: Home > Modify > Move **Toolbar:** Modify > Move
Command: MOVE/M

The **Move** tool is used to move one or more objects from their current location to a new location without changing their size or orientation.

COPYING SKETCHED OBJECTS

Ribbon: Home > Modify > Copy **Toolbar:** Modify > Copy
Command: COPY/CO

The **Copy** tool is used to copy an existing object. This tool is used to make the copies of the selected objects and place them at the specified location.

PASTING CONTENTS FROM THE CLIPBOARD

Ribbon: Home > Clipboard > Paste > Paste as Block **Command:** PASTEBLOCK

The **Paste as Block** tool is used to paste the contents of the Clipboard into a new drawing or in the same drawing at a new location. You can also invoke the **PASTEBLOCK** command from the shortcut menu by right-clicking in the drawing area and then choosing **Clipboard > Paste as Block**.

PASTING CONTENTS USING THE ORIGINAL COORDINATES

Ribbon: Home > Clipboard > Paste > Paste to Original Coordinates
Command: PASTEORIG

The **Paste to Original Coordinates** tool is used to paste the contents of the Clipboard into a new drawing by using coordinates from the original drawing. You can invoke the **PASTEORIG** command from the shortcut menu by right-clicking in the drawing area and then choosing **Clipboard > Paste to Original Coordinates**. Note that this command will be available only when the Clipboard contains AutoCAD data from a drawing other than the current drawing.

OFFSETTING SKETCHED OBJECTS

Ribbon: Home > Modify > Offset **Toolbar:** Modify > Offset
Command: OFFSET/O

You can use the **Offset** tool to draw parallel lines, polylines, concentric circles, arcs, curves, and so on. However, you can offset only one entity at a time. While offsetting an object, you need to specify the offset distance and the side to offset, or specify the distance through which the selected object has to be offset. Depending on the side to offset, the resulting object will be smaller or larger than the original object. For example, while offsetting a circle if the offset side is toward the inner side of the perimeter, the resulting circle will be smaller than the original one.

ROTATING SKETCHED OBJECTS

Ribbon: Home > Modify > Rotate **Toolbar:** Modify > Rotate
Command: ROTATE/RO

While creating designs, sometimes you have to rotate an object or a group of objects. You can accomplish this by using the **Rotate** tool. On invoking this tool, you will be prompted to select the objects and the base point about which the selected objects will be rotated. You should be careful in selecting the base point if the base point is not located on the known object. After you specify the base point, you need to enter the rotation angle. By default, a positive angle results in a counterclockwise rotation, whereas a negative angle results in a clockwise rotation. The **Rotate** tool can also be invoked from the shortcut menu by selecting an object and right-clicking in the drawing area and choosing the **Rotate** tool from the shortcut menu.

SCALING THE SKETCHED OBJECTS

Ribbon: Home > Modify > Scale **Toolbar:** Modify > Scale
Command: SCALE/SC

Sometimes you need to change the size of objects in a drawing. You can do so by using the **Scale** tool. This tool dynamically enlarges or shrinks a selected object about a base point keeping the aspect ratio of the object constant. This means that the size of the object will be increased or reduced equally in the X, Y, and Z directions. The dynamic scaling property allows you to view the object as it is being scaled. Another advantage of this tool is that if you have dimensioned the drawing, they will also change accordingly. You can also invoke the **Scale** tool from the shortcut menu by right-clicking in the drawing area after selecting an object and choosing the **Scale** tool.

FILLETING THE SKETCHES

Ribbon: Home > Modify > Fillet drop-down > Fillet **Toolbar:** Modify > Fillet
Command: FILLET/F

The edges in a model are generally filleted to reduce the area of stress concentration. The **Fillet** tool helps you form round corners between two entities that form a sharp vertex. As a result, a smooth round arc is created that connects the two objects. A fillet can also be created between two intersecting or parallel lines as well as non-intersecting and nonparallel lines, arcs, polylines, xlines, rays, splines, circles, and true ellipses. The fillet arc created will be tangent to both the selected entities. The default fillet radius is 0.0000. Therefore, after invoking this tool, you first need to specify the radius value.

CHAMFERING THE SKETCHES

Ribbon: Home > Modify > Fillet drop-down > Chamfer
Toolbar: Modify > Chamfer **Command:** CHAMFER/CHA

Chamfering the sharp corners is another method of reducing the areas of stress concentration in a model. Chamfering is defined as the process by which the sharp edges or corners are beveled. The size of a chamfer depends on its distance from the corner. If a chamfer is equidistant from the corner in both directions, it is a 45-degree chamfer. A chamfer can be drawn between two lines that may or may not intersect. This tool also works on a single polyline. In AutoCAD, the chamfers are created by defining two distances or by defining one distance and the chamfer angle.

BLENDING THE CURVES

Ribbon: Home > Modify > Fillet drop-down > Blend Curves
Command: BLEND

In AutoCAD, you can create a smooth or tangent continuous spline between the endpoints of two existing curves. The existing curves can be arcs, lines, helixes, open polylines, or open splines. To create a smooth curve between two open curves, choose the **Blend Curves** tool from the **Fillet** drop-down in the **Modify** panel; you will be prompted to select the curves one after the other. Select two curves; a blend curve will be created between the selected curves. The length of the selected curves does not change.

TRIMMING THE SKETCHED OBJECTS

Ribbon: Home > Modify > Trim drop-down > Trim **Toolbar:** Modify > Trim
Command: TRIM/TR

While creating a design, you may need to remove the unwanted and extended edges. Breaking individual objects takes time if you are working on a complex design with many objects. In such cases, you can use the **Trim** tool. This tool is used to trim the objects that extend beyond a required point of intersection. When you invoke this tool from the **Trim** drop-down in the **Modify** panel, you will be prompted to select the cutting edges or boundaries. These edges can be lines, polylines, circles, arcs, ellipses, xlines, rays, splines, text, blocks, or even viewports. There can be more than one cutting edge and you can use any selection method to select them. After the cutting edge or edges are selected, you must select each object to be trimmed. An object can be both a cutting edge and an object to be trimmed. You can trim lines, circles, arcs, polylines, splines, ellipses, xlines, and rays.

EXTENDING THE SKETCHED OBJECTS

Ribbon: Home > Modify > Trim drop-down > Extend
Toolbar: Modify > Extend **Command:** EXTEND/EX

The **Extend** tool may be considered as the opposite of the **Trim** tool. You can extend lines, polylines, rays, and arcs to connect to other objects by using the **Extend** tool. However, you cannot extend closed loops. The command prompt of the **Extend** tool is similar to that of the **Trim** tool. You are required to select the boundary edges first. The boundary edges are those objects that the selected lines or arcs extend to meet. These edges can be lines, polylines, circles, arcs, ellipses, xlines, rays, splines, text, blocks, or even viewports.

STRETCHING THE SKETCHED OBJECTS

Ribbon: Home > Modify > Stretch **Toolbar:** Modify > Stretch
Command: STRETCH/S

This tool is used to lengthen objects, shorten them, and alter their shapes, refer to Figure 5-1. To invoke this tool, choose the **Stretch** tool from the **Modify** panel; you will be prompted to select objects. Use the **Crossing** or **CPolygon** selection method to select the objects to be stretched. After selecting the objects, you will be prompted to specify the base point of displacement. Select the portion of the object that needs to be stretched; you will be prompted to specify the base point. After specifying the base point, you will be prompted to specify the second point of displacement. Specify the new location; the object will lengthen or shorten. Figure 5-1 illustrates the usage of a crossing selection to simultaneously select two angled lines and then stretch their right ends.

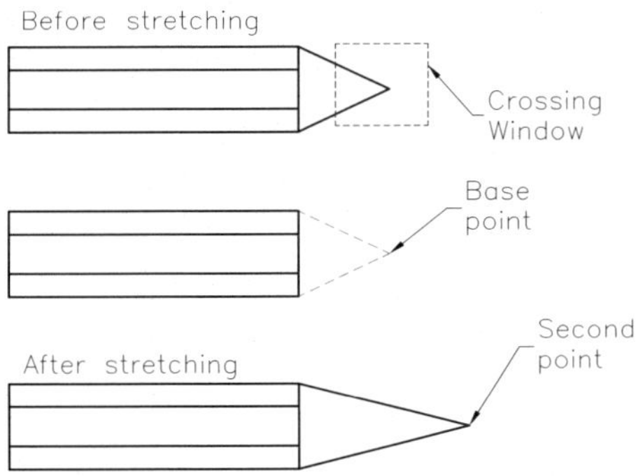

Figure 5-1 Stretching the entities

Note
The regions and solids cannot be stretched. If you select them, they will move instead of getting stretched.

LENGTHENING THE SKETCHED OBJECTS

Ribbon: Home > Modify > Lengthen **Command:** LENGTHEN/LEN

Like the **Trim** and **Extend** tools, the **Lengthen** tool can also be used to extend or shorten lines, polylines, elliptical arcs, and arcs. This tool has several options that allow you to change the length of objects by dynamically dragging the object endpoint, entering the delta value, entering the percentage value, or entering the total length of the object. This tool also allows the repeated selection of the objects for editing.

ARRAYING THE SKETCHED OBJECTS

Ribbon: Home > Modify > Array **Command:** ARRAY/AR

The arrays can be created by using the **Array** drop-down, refer to Figure 5-2. You can use this drop-down to create a rectangular, path, or polar array. You can also enter **ARRAY** in the Command Bar to invoke the options for creating an array. The different types of arrays that can be created using the **Array** drop-down are discussed next.

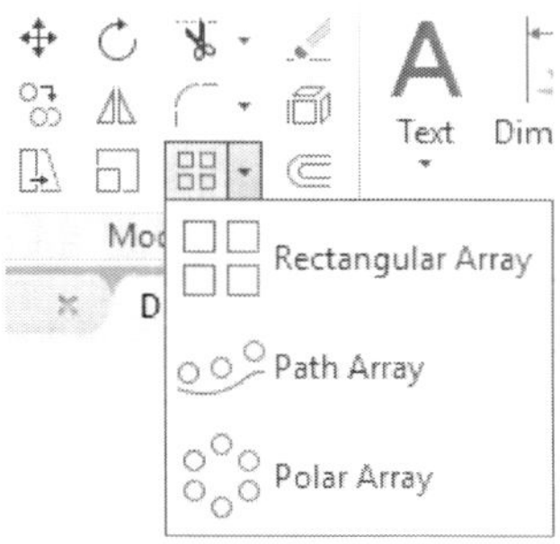

Figure 5-2 *Tools in the* ***Array*** *drop-down*

Rectangular Array

Ribbon: Home > Modify > Array drop-down > Rectangular Array
Toolbar: Modify > Array drop-down > Rectangular Array
Command: ARRAYRECT

A rectangular array is formed by making copies of the selected object along the *X* and *Y* directions of an imaginary rectangle (along rows and columns). To create a rectangular array, choose the **Rectangular Array** tool from the **Modify** panel; you will be prompted to select objects. Select the objects to be arrayed and press ENTER; you will be prompted to select the grip to edit the array. Select the grip and click to specify the spacing for the array, as required. Also, you can enter the value of spacing. Next, press ENTER or X; the array of the selected object will be created.

Polar Array

Ribbon: Home > Modify > Array drop-down > Polar Array
Toolbar: Modify > Array drop-down > Polar Array **Command:** ARRAYPOLAR

Polar array is an arrangement of objects in circular pattern around a point. A polar array can be created by choosing the **Polar Array** tool from the **Array** drop-down of the **Modify** panel, refer to Figure 5-2. When you choose this tool, you will be prompted to select objects. Select objects to be arrayed; you will be prompted to specify the center point of the array. Select the center point of the array; you will be prompted to select grip to edit array. Press ENTER or X to exit the tool.

Path Array

Ribbon: Home > Modify > Array drop-down > Path Array
Toolbar: Modify > Array drop-down > Path Array **Command:** ARRAYPATH

In AutoCAD, you can create an array of objects along a path called path arrays. The path can be a line, polyline, circle, helix, and so on. To create a path array, choose the **Path Array** tool from the **Modify** panel, refer to Figure 5-2. On doing so, you will be prompted to select objects. Select the objects to be arrayed and press ENTER; you will be prompted to select the path curve along which the object will be arrayed. Select the path curve; the preview of the path array will be displayed and you will be prompted to select the grip to edit array. Choose the desired option at the command prompt and press ENTER to accept or X to exit the tool.

MIRRORING THE SKETCHED OBJECTS

Ribbon: Home > Modify > Mirror **Toolbar:** Modify > Mirror
Command: MIRROR/MI

The **Mirror** tool is used to create a mirrored copy of the selected objects. The objects can be mirrored at any angle. This tool is helpful in drawing symmetrical figures. On invoking this tool, you will be prompted to select objects. On selecting the objects to be mirrored, you will be prompted to enter the first point of the mirror line and the second point of the mirror line. On selecting the first point of the mirror line, a preview of the mirrored objects will be displayed. Next, you need to specify the second endpoint of the mirror line, as shown in Figure 5-3. On selecting the second endpoint, you will be prompted to specify whether you want to delete the source object or not. Enter **Yes** to delete the source object and **No** to retain the source object, refer to Figure 5-4.

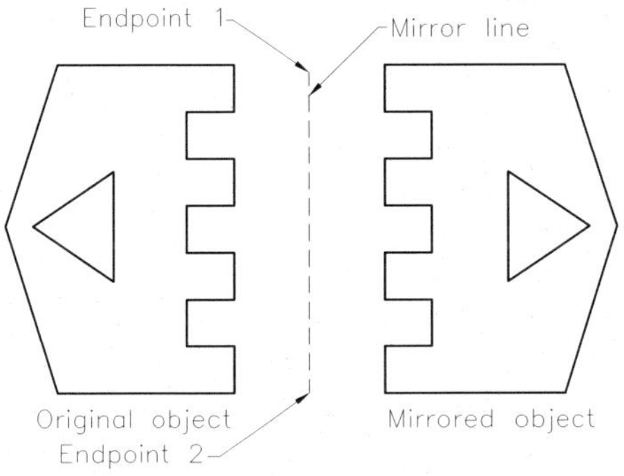

Figure 5-3 *Creating a mirror image of an object by using the* ***Mirror*** *tool*

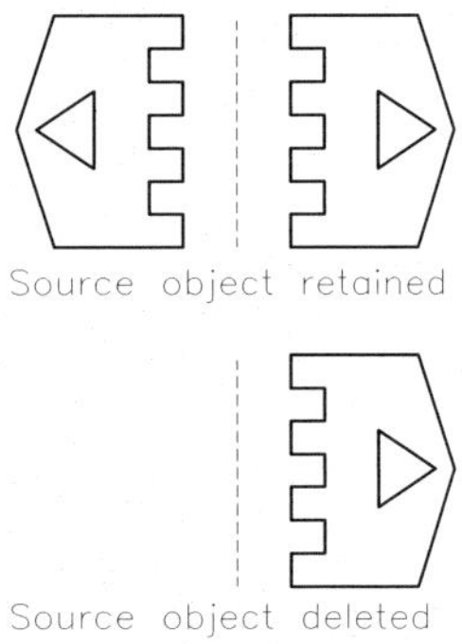

Figure 5-4 *Retaining and deleting source objects after mirroring*

BREAKING THE SKETCHED OBJECTS

Ribbon: Home > Modify > Break at Point, Break **Command:** BREAK/BR
Toolbar: Modify > Break at Point, Break

The **Break** tool breaks an existing object into two parts or erases a portion of an object. This tool can be used to remove a part of the selected objects or it can be used to break objects such as lines, arcs, circles, ellipses, xlines, rays, splines, and polylines. You can break the objects using the methods discussed next.

Break at Point

Choose the **Break at Point** tool from the **Modify** panel to break an object into two parts by specifying a breakpoint. On choosing this tool, you will be prompted to select the object to be broken. Once you have selected the object, you will be prompted to specify the first break point. Specify the first break point on the object; the object will be broken.

Break

To break an object by removing a portion of the object between two selected points, choose the **Break** tool from the **Modify** panel; you will be prompted to select the object. Select the object; you will be prompted to select the second break point. Select a point on the object; the portion of the object between the two selected points will be removed. Note that the point at which you select an object becomes the first break point.

JOINING THE SKETCHED OBJECTS

Ribbon: Home > Modify > Join **Command:** JOIN/J

The **Join** tool is used to join two or more collinear lines or arcs lying on the same imaginary circle. The lines or arcs can have gap between them. This tool can also be used to join two or more polylines or splines, but they should be on the same plane and should not have any gap between them.

Tutorial 1 Rectangular and Polar Arrays

In this tutorial, you will create the drawing of an end plate. The dimensions of the end plate are shown in Figure 5-5.

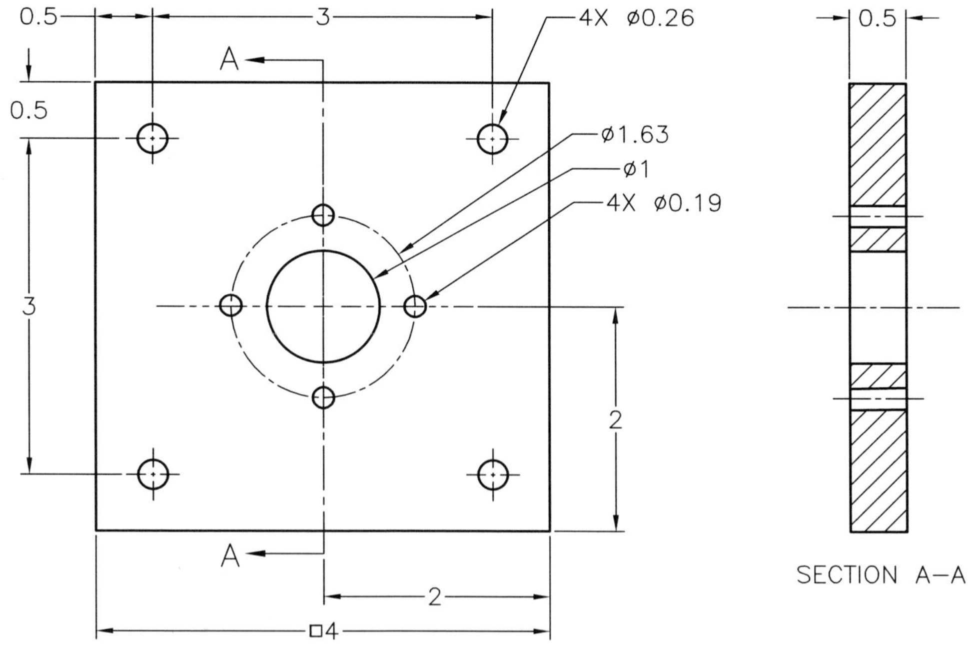

Figure 5-5 *Drawing for Tutorial 1*

First, you need to create a rectangle by using the **Rectangle** tool.

1. Start a new drawing file in the **Drafting & Annotation** workspace.

2. Choose the **Rectangle** tool from **Home > Draw > Rectangle** drop-down and draw a rectangle of **4** units in length and breadth each. The prompt sequence to draw the rectangle is:

   *Choose the **Rectangle** tool from the **Draw** panel*
   Specify first corner point or [Chamfer/Elevation/Fillet/Thickness/Width]: **0,0** Enter
   Specify other corner point or [Area/Dimensions/Rotation]: **@4,4** Enter

After creating the rectangle, you need to create three circles of different radii. These circles can be created by using the **Center, Radius** tool. You need to use the **From** object snap mode to create the third circle.

3. Choose the **Center, Radius** tool from **Home > Draw > Circle** drop-down and draw a circle of diameter **1** unit, as shown in Figure 5-6. The prompt sequence to draw the circle is:

   *Choose the **Center, Radius** tool from the **Draw** panel*
   Specify center point for circle or [3P/2P/Ttr (tan tan radius)]: **2,2** Enter
   Specify radius of circle or [Diameter] <current>: **D** Enter
   Specify diameter of circle <current>: **1** Enter

4. Press ENTER to invoke the **Center, Radius** tool again and draw a circle of diameter **0.26** unit, as shown in Figure 5-7. You can also invoke the **Center, Radius** tool from the **Ribbon**. The prompt sequence to draw the circle is:

   *Choose the **Center, Radius** tool from the **Draw** panel*
   Specify center point for circle or [3P/2P/Ttr (tan tan radius)]: **0.5**, **0.5** Enter
   Specify radius of circle or [Diameter] <current>: **D** Enter
   Specify diameter of circle <current>: **0.26** Enter

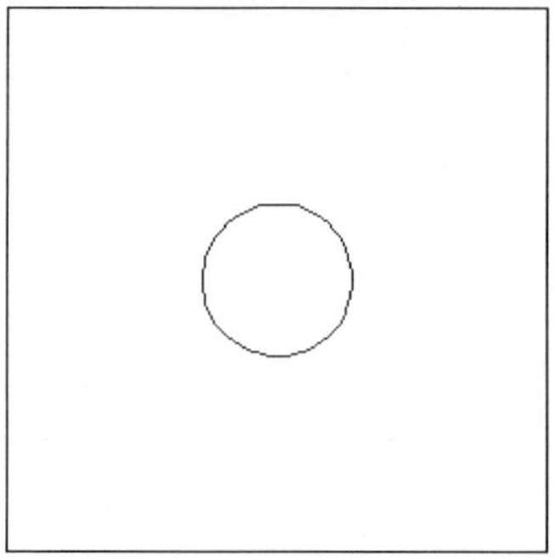

Figure 5-6 *Rectangle and circle created*

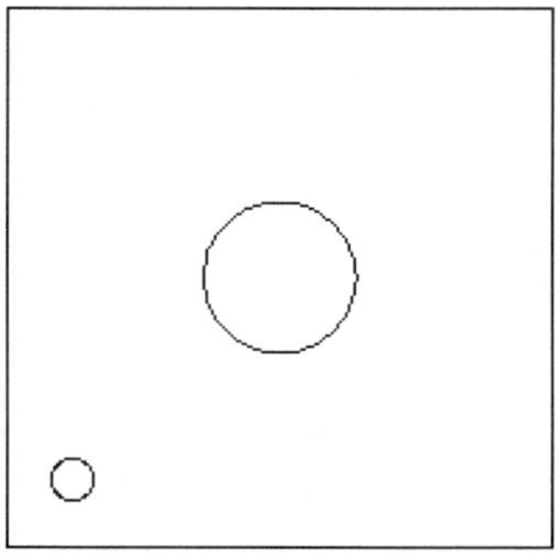

Figure 5-7 *Circle of diameter **0.26** unit created*

5. Again, press ENTER to invoke the **Center, Radius** tool. You can also invoke this tool from the **Draw** panel. Next, press SHIFT and then right-click to invoke a shortcut menu.

6. Choose the **From** option from the shortcut menu and then draw the circle, as shown in Figure 5-8 by following the prompt given next.

   *Choose the **Center, Radius** tool from the **Draw** panel*
   Specify center point for circle or [3P/2P/Ttr (tan tan radius)]: _from Base point: *Click the*

center of the circle of diameter 1 unit
_from Base point: <Offset>: **@-0.8125,0** Enter
Specify radius of circle or [Diameter] <current>: **D** Enter
Specify diameter of circle <current>: **0.19** Enter

After creating all circles, you need to create the pattern of two circles created last. The first array will be a rectangular array and second array will be a polar array. These arrays will be created by using the **Rectangular Array** and **Polar Array** tools.

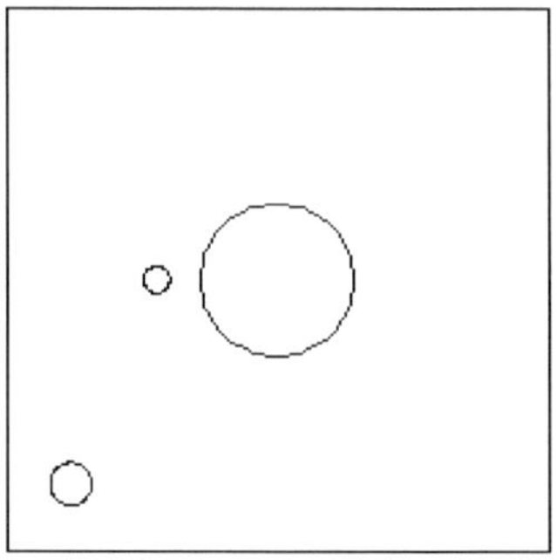

Figure 5-8 *Top view of the model*

7. Choose the **Rectangular Array** tool from **Home > Modify > Array** drop-down and follow the command prompt given next to create a rectangular array of the circle of diameter **0.26** unit shown in Figure 5-9.

*Choose the **Rectangular Array** tool from the **Modify** panel*
Select objects: *Select the circle of diameter 0.26 units and press ENTER*
Type = Rectangular Associative = Yes
Select grip to edit array or [ASsociative/Base point/COUnt/Spacing/COLumns/Rows/Levels/eXit]<eXit>: COU Enter
Enter the number of columns or [Expression] <4>: 2 Enter
Enter the number of rows or [Expression] <3>: 2 Enter
Select grip to edit array or [ASsociative/Base point/COUnt/Spacing/COLumns/Rows/Levels/eXit]<eXit>: S Enter
Specify the distance between columns or [Unit cell]<current>: 3 Enter
Specify the distance between rows <current>: 3 Enter
Select grip to edit array or [ASsociative/Base point/COUnt/Spacing/COLumns/Rows/Levels/eXit]<eXit>: X Enter

8. Next, choose the **Polar Array** tool from **Draw > Modify > Array** drop-down and follow the prompt given next to create a polar array of the circle of diameter **0.19** unit shown in Figure 5-10.

*Choose the **Polar Array** tool from the **Modify** panel*
Select objects: *Select the circle of diameter **0.19** unit and press ENTER*
Type = Polar Associative = Yes
Specify center point of array or [Base point/Axis of rotation]: *Click at the center point of circle of diameter 1 unit*
Select grip to edit array or [ASsociative/Base point/Items/Angle between/Fill angle/ROWs/Levels/ROTate items/eXit]<eXit>: I Enter
Enter number of items in array or [Expression] <current>: **4** Enter
Select grip to edit array or [ASsociative/Base point/Items/Angle between/Fill angle/ROWs/Levels/ROTate items/eXit]<eXit>: F Enter
Specify the angle to fill (+=ccw, -=cw) or [EXpression] <360>: **360** Enter
Press Enter to accept or [ASsociative/Base point/Items/Angle between/Fill angle/ROWs/Levels/ROTate items/eXit]<eXit>: **X** Enter

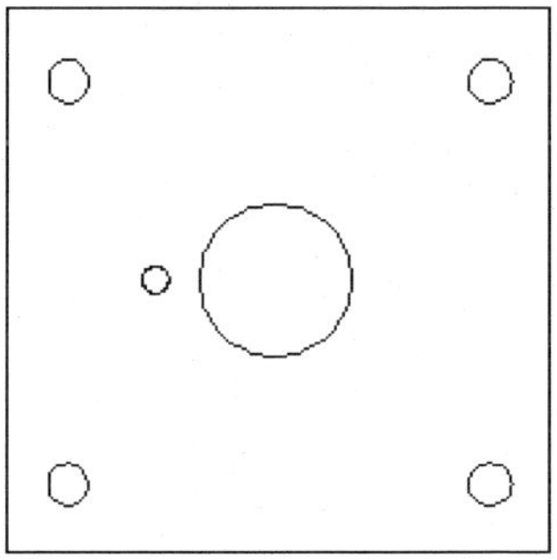

Figure 5-9 *Rectangular array created*

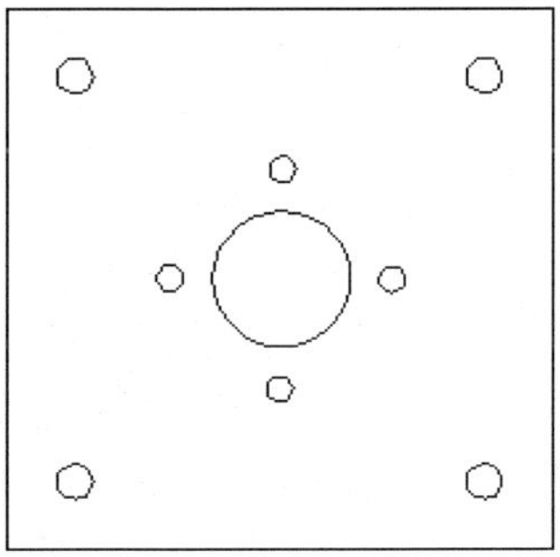

Figure 5-10 *Polar array created*

EXERCISES

Exercise 1

Create the drawing shown in Figure 5-11 and save it.

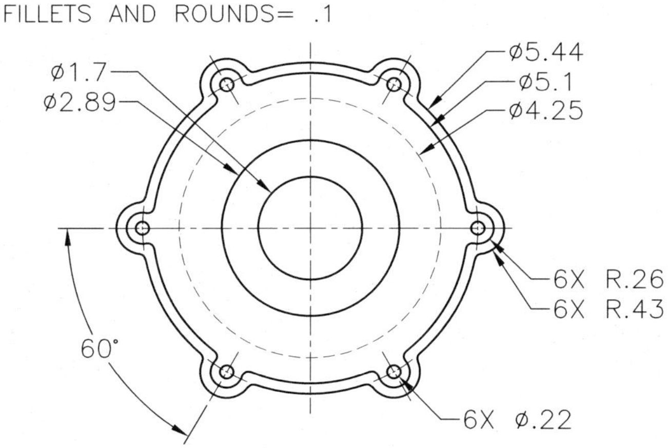

Figure 5-11 *Drawing for Problem-Solving Exercise 1*

Exercise 2

Draw the dining table with chairs, as shown in Figure 5-12, and then save the drawing.

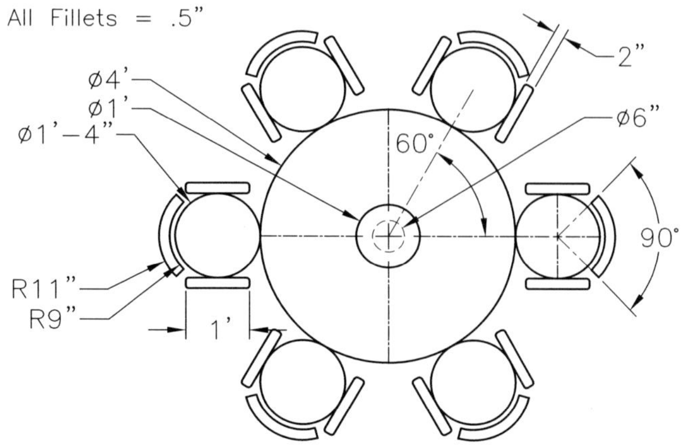

Figure 5-12 *Drawing for Problem-Solving Exercise 2*

Chapter 6

Basic Dimensioning, Geometric Dimensioning, and Tolerancing

Learning Objectives

After completing this chapter, you will be able to:

- *Understand the need of dimensioning in drawings*
- *Understand fundamental dimensioning terms*
- *Apply the associative and annotative dimensioning*
- *Use the Quick Dimension option for quick dimensioning*
- *Create various types of dimensions in a drawing*
- *Use geometric tolerancing*

NEED FOR DIMENSIONING

To make designs more informative and practical, a drawing must convey more than just the graphic picture of a product. To manufacture an object, the drawing of that object must contain size descriptions such as the length, width, height, angle, radius, diameter, and location of features. These informations are added to the drawing by dimensioning. Some drawings also require information about tolerances with the size of features. The information conveyed through dimensioning are vital and often as important as the drawing itself. With the advancement in computer-aided design/drafting and computer-aided manufacturing, it has become mandatory to draw part to actual size so that dimensions reflect the actual size of features. At times, it may not be necessary to draw the object of the same size as the actual object would be, but it is absolutely essential that the dimensions be accurate. Incorrect dimensions will lead to manufacturing errors.

DIMENSIONING IN AutoCAD

The objects that can be dimensioned in AutoCAD range from straight lines to arcs. The dimensioning tools provided by AutoCAD can be classified into four categories:

Dimension Drawing tools	Dimension Style tools
Dimension Editing tools	Dimension Utility tools

ASSOCIATIVE DIMENSIONS

The Associative dimensioning is a method of dimensioning in which the dimension is associated with the object that is dimensioned. In other words, the dimension is influenced by the changes in the size of the object.

DEFINITION POINTS

Definition points are the points drawn at the positions used to generate a dimension object. The definition points are used by the dimensions to control their updating and rescaling. AutoCAD draws these points on a special layer called **Defpoints**.

ANNOTATIVE DIMENSIONS

When all the elements of a dimension such as text, spacing, and arrows get scaled according to the specified annotation scale, it is known as Annotative Dimension. They are created in the drawing by assigning annotative dimension styles to them.

DIMENSIONING A NUMBER OF OBJECTS TOGETHER

Ribbon: Annotate > Dimensions > Quick	**Command:** QDIM
Menu Bar: Dimension > Quick Dimension	**Toolbar:** Dimension > Quick Dimension

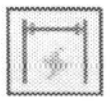

The **Quick Dimension** tool is used to dimension a number of objects at the same time. It also helps you to quickly edit dimension arrangements already existing in drawing and also creates new dimension arrangements. It is especially useful while creating a series of baseline or continuous dimensions. It also allows you to dimension multiple arcs and circles at the same time.

CREATING LINEAR DIMENSIONS

Ribbon: Annotate > Dimensions > Dimension drop-down > Linear
Menu Bar: Dimension > Linear **Toolbar:** Dimension > Linear
Command: DIMLIN or DIMLINEAR

Linear dimensioning is used to measure the horizontal or vertical distance between two points. You can directly select the object to dimension or select two points. The points can be any two points in the space, endpoints of an arc or line, or any set of points that can be identified. The prompt sequence that will follow when you choose the **Linear** tool is given next.

Specify first extension line origin or <select object>: Enter
Select object to dimension: *Select the object.*
Specify dimension line location or
[Mtext/Text/Angle/Horizontal/Vertical/Rotated]: *Select a point to locate the position of the dimension.*

Tutorial 1 Horizontal Dimension

In this tutorial, you will use linear dimensioning to dimension a horizontal line of 4 units length. The dimensioning will be done first by selecting the object and later on by specifying the first and second extension line origins. Using the **Text Editor** tab, modify the default text such that the dimension is underlined.

Selecting the Object

1. Start a new file in the **Drafting & Annotation** workspace and draw a line of 4 units length.

2. Choose the **Linear** tool from **Annotate** > **Dimensions** > **Dimension** drop-down; you will be prompted to specify the extension line origin or the object. The prompt sequence to apply the linear dimension is as follows:

Specify first extension line origin or <select object>: Enter
Select object to dimension: *Select the line.*
Specify dimension line location or
[Mtext/Text/Angle/Horizontal/Vertical/Rotated]: **M** Enter

*The **Text Editor** tab will be displayed, as shown in Figure 6-1. Select the default dimension value and then choose the **Underline** button from the **Formatting** panel of the **Text Editor** tab of the **Ribbon** to underline the text. Click anywhere in the drawing area to exit the **Text Editor** tab.*

Specify dimension line location or
[Mtext/Text/Angle/Horizontal/Vertical/Rotated]: *Place the dimension.*
Dimension text = 4.0000

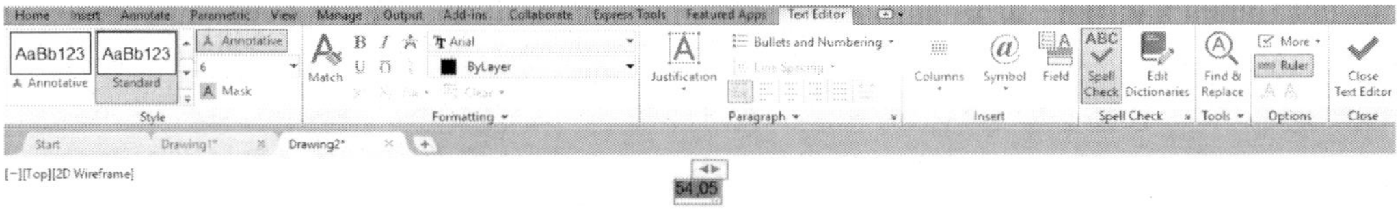

*Figure 6-1 The **Text Editor** tab*

Specifying Extension Line Origins

1. Choose the **Linear** tool from **Annotate** > **Dimensions** > **Dimension** drop-down. The prompt sequence is as follows:

   Specify first extension line origin or <select object>: *Select the first endpoint of the line using the* ***Endpoint*** *object snap, refer to Figure 6-2.*

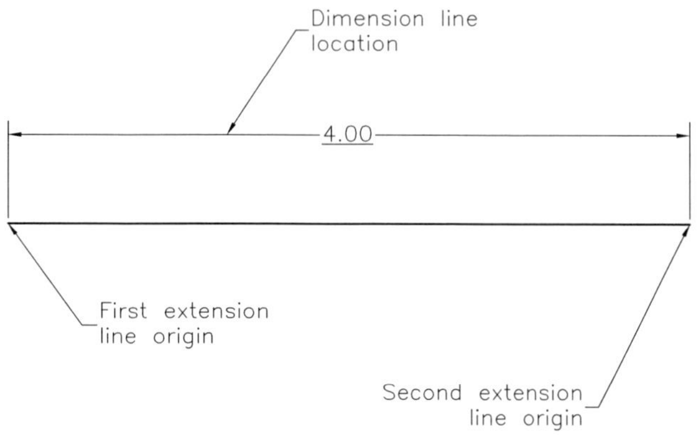

Figure 6-2 *Line for Tutorial 1*

   Specify second extension line origin: *Select the second endpoint of the line using the* ***Endpoint*** *object snap, refer to Figure 6-2.*
   Specify dimension line location or
   [Mtext/Text/Angle/Horizontal/Vertical/Rotated]: **M** Enter
   Select the text and then choose the ***Underline*** *button from the* ***Ribbon*** *to underline the text in the* ***Text Editor****. Click anywhere in the drawing area.*
   Specify dimension line location or
   [Mtext/Text/Angle/Horizontal/Vertical/Rotated]: *Place the dimension.*
   Dimension text = 4.00

CREATING ALIGNED DIMENSIONS

Ribbon: Annotate > Dimensions > Dimension drop-down > Aligned
Menu Bar: Dimension > Aligned **Toolbar:** Dimension > Aligned
Command: DIMALI or DIMALIGNED

Generally, the drawing consists of various objects that are neither parallel to the *X* axis nor to the *Y* axis. Dimensioning of such objects can be done using aligned dimensioning. In horizontal or vertical dimensioning, you can only measure the shortest distance from the first extension line origin to the second extension line origin along the horizontal or vertical axis, respectively whereas with the help of aligned dimensioning, you can measure the true aligned distance between two points. The function of the **Aligned** tool is similar to that of the other linear dimensioning tools. The dimension created with the **Aligned** tool is parallel to the object being dimensioned.

CREATING ARC LENGTH DIMENSIONS

Ribbon: Annotate > Dimensions > Dimension drop-down > Arc Length
Menu Bar: Dimension > Arc Length **Toolbar:** Dimension > Arc Length
Command: DIMARC

The Arc Length dimensioning is used to dimension the length of an arc or the polyline arc segment. You are required to select an arc or a polyline arc segment and the dimension location. Choose the **Arc Length** tool in the **Dimensions** panel. The prompt sequence that will follow after choosing this tool is given next.

Select arc or polyline arc segment: *Select arc or polyline arc segment to dimension.*
Specify arc length dimension location, or [Mtext/Text/Angle/Partial/Leader]: *Specify the location for the dimension line.*

CREATING BASELINE DIMENSIONS

Ribbon: Annotate > Dimensions > Continue drop-down > Baseline
Menu Bar: Dimension > Baseline **Toolbar:** Dimension > Baseline
Command: DIMBASE or DIMBASELINE

Sometimes in manufacturing, you may want to locate different points and features of a part with reference to a fixed point (base point or reference point). This can be accomplished by using the Baseline dimensioning. To invoke the Baseline dimension, choose the **Baseline** tool from the **Continue** drop-down of the **Dimensions** panel. Using this tool, you can continue a linear dimension from the first extension line origin of the first dimension to the dimension point. The new dimension line is automatically offset by a fixed amount to avoid overlapping of the dimension lines. This has to be kept in mind that there must be already existing a linear, ordinate, or angular associative dimension to use the Baseline dimensions.

CREATING CONTINUED DIMENSIONS

Ribbon: Annotate > Dimensions > Continue drop-down > Continue
Menu Bar: Dimension > Continue **Toolbar:** Dimension > Continue
Command: DIMCONT or DIMCONTINUE

Using the **Continue** tool, you can continue a linear dimension from the second extension line of the previous dimension. This is also called as Chained or Incremental dimensioning. Note that there must be existing linear, ordinate, or angular associative dimension to use the Continue dimensions.

CREATING ANGULAR DIMENSIONS

Ribbon: Annotate > Dimensions > Dimension drop-down > Angular
Menu Bar: Dimension > Angular **Toolbar:** Dimension > Angular
Command: DIMANG or DIMANGULAR

The Angular dimensioning is used for applying angular dimension to an entity. The **Angular** tool is used to generate a dimension arc (dimension line in the shape of an arc with arrowheads at both ends) to indicate the angle between two nonparallel lines. This tool can also be used to dimension the vertex and two other points, a circle with another point, or the angle of an arc. For every set of points, there exists one acute angle and one obtuse angle (inner and outer angles). If you specify the dimension arc location between the two points, you will get the acute angle; if you specify it outside the two points, you will get the obtuse angle.

CREATING DIAMETER DIMENSIONS

Ribbon: Annotate > Dimensions > Dimension drop-down > Diameter
Menu Bar: Dimension > Diameter **Toolbar:** Dimension > Diameter
Command: DIMDIA or DIMDIAMETER

Diameter dimensioning is used to dimension a circle or an arc. Here, the measurement is done between two diametrically opposite points on the circumference of the circle or the arc. The dimension text generated by AutoCAD begins with the Ø symbol to indicate a diameter dimension.

CREATING JOGGED DIMENSIONS

Ribbon: Annotate > Dimensions > Dimension drop-down > Jogged
Menu Bar: Dimension > Jogged **Toolbar:** Dimension > Jogged
Command: DIMJOGGED

The necessity of the Jogged dimension arises because of the space constraint. Also, the Jogged dimension is used when you want to avoid the merging of the dimension line with other dimensions. Also, there are instances when it is not possible to show the center of the circle in the sheet. In such situations, the jogged dimensions are used. The jogged dimensions can be added using the **Jogged** tool. Note that with this tool, you can add only jogged radius dimensions. To add jogged dimensions, invoke the **Jogged** tool from the **Dimensions** panel and select an arc or a circle from the drawing area; you will be prompted to select the center location override. Specify a new location to override the existing center. This point will also become the start point of the dimension line. Next, specify the location of the dimension line and the jog.

CREATING RADIUS DIMENSIONS

Ribbon: Annotate > Dimensions > Dimension drop-down > Radius
Menu Bar: Dimension > Radius **Toolbar:** Dimension > Radius
Command: DIMRAD or DIMRADIUS

The Radius dimensioning is used to dimension a circle or an arc. Radius and diameter dimensioning are similar; the only difference is that instead of the diameter line, a radius line is drawn (half of the diameter line), which is measured from the center to any point on the circumference. The dimension text generated by AutoCAD is preceded by the letter R to indicate a radius dimension. If you want to use the default dimension text (dimension text generated automatically by AutoCAD), simply specify a point to position the dimension at the **Specify dimension line location or [Mtext/Text/Angle]** prompt.

GEOMETRIC DIMENSIONING AND TOLERANCING

One of the most important parts of the design process is assigning the dimensions and tolerances to parts, since every part is manufactured from the dimensions given in the drawing. Therefore, every designer must understand and have a thorough knowledge of the standard practices used in the industry to make sure that the information given on the drawing is correct and can be understood by other people. Tolerancing is equally important, especially in the assembled parts. Tolerances and fits determine how the parts will fit. Incorrect tolerances could result in a product that is not usable. In addition to dimensioning and tolerancing, the function and the relationship that exists between the mating parts is important if the part is to perform the way it was designed. This aspect of the design process is addressed by geometric dimensioning and tolerancing, generally known as GD & T.

ADDING GEOMETRIC TOLERANCE

Ribbon: Annotate > Dimensions > Tolerance **Toolbar:** Dimension > Tolerance
Menu Bar: Dimension > Tolerance **Command:** TOLERANCE/TOL

Geometric tolerance displays the deviations of profile, orientation, form, location, and runout of a feature. In AutoCAD, geometrical tolerancing is displayed by feature control frames. The frames contain all information about tolerances for a single dimension. To display feature control frames with various tolerancing parameters, you need to enter specifications in the **Geometric Tolerance** dialog box, as shown in Figure 6-3. You can invoke the **Geometric Tolerance** dialog box by choosing the **Tolerance** tool from the extended **Dimensions** panel.

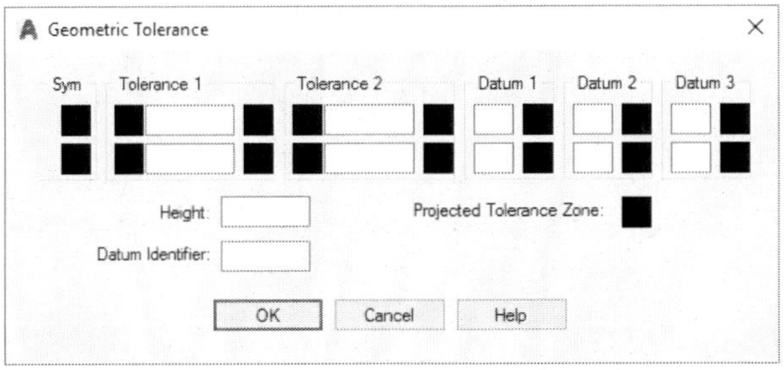

*Figure 6-3 The **Geometric Tolerance** dialog box*

Tutorial 2 *Tolerance*

In this tutorial, you will create a feature control frame to define the perpendicularity specification, as shown in Figure 6-4.

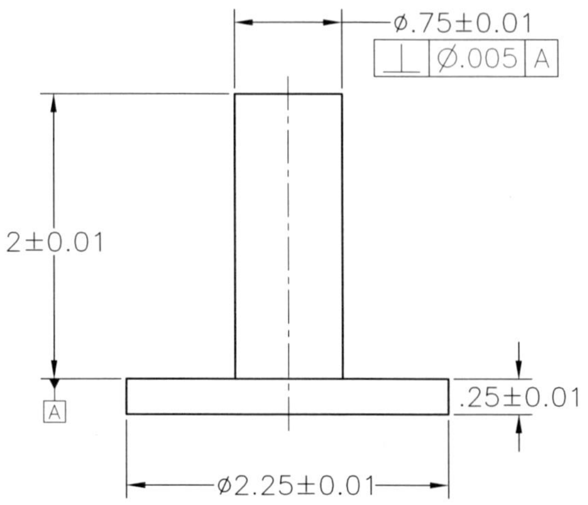

Figure 6-4 *Drawing for Tutorial 2*

1. Choose the **Tolerance** tool from the **Dimensions** panel of the **Annotate** tab to display the **Geometric Tolerance** dialog box. Choose the upper box from the **Sym** area to display the **Symbol** dialog box. Select the perpendicularity symbol. This symbol will be displayed in the **Sym** area.

2. Select the box on the left of the upper edit box under the **Tolerance 1** area; the diameter symbol will appear to denote a cylindrical tolerance zone.

3. Enter **.005** in the upper edit box under the **Tolerance 1** area in the **Geometric Tolerance** dialog box.

4. Enter **A** in the edit box under the **Datum 1** area. Choose the **OK** button to accept the changes made in the **Geometric Tolerance** dialog box.

5. The **Enter tolerance location** prompt is displayed in the command line area and the feature control frame is attached to the cursor at its middle left point. Select a point to insert the frame.

6. To attach the datum symbol, invoke the **Leader Settings** dialog box and select the **Tolerance** radio button as explained in step 1.

7. Choose the **Leader Line & Arrow** tab and select the **Datum triangle filled** option from the drop-down list in the **Arrowhead** area.

8. Set the number of points in the leader to **2** in the **Maximum** spinner of **Number of Points** area.

9. Choose the **OK** button to return to the command line. Specify the first and second leader points, refer to Figure 6-4. On specifying the second point, the **Geometric Tolerance** dialog box will be displayed.

10. Enter **A** in the **Datum Identifier** edit box. Choose the **OK** button to exit the **Geometric Tolerance** dialog box. On doing so, the datum symbol will be displayed, refer to Figure 6-4.

EXERCISES

Exercise 1

Draw the object shown in Figure 6-5 and then dimension it. Save the drawing as *DIMEXR1*.

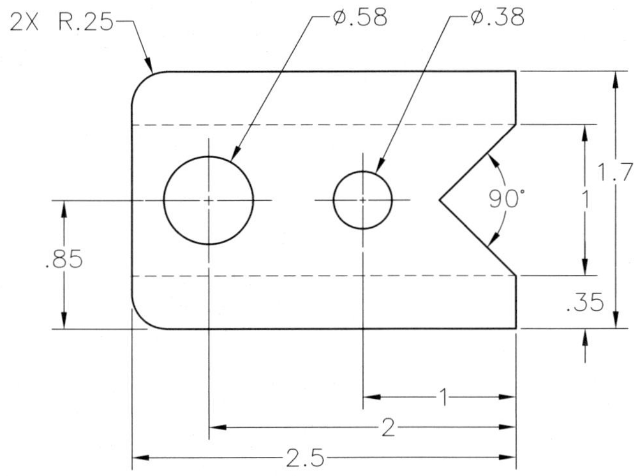

Figure 6-5 *Drawing for Exercise 1*

Exercise 2

Draw and dimension the object shown in Figure 6-6. Save the drawing as *DIMEXR2*.

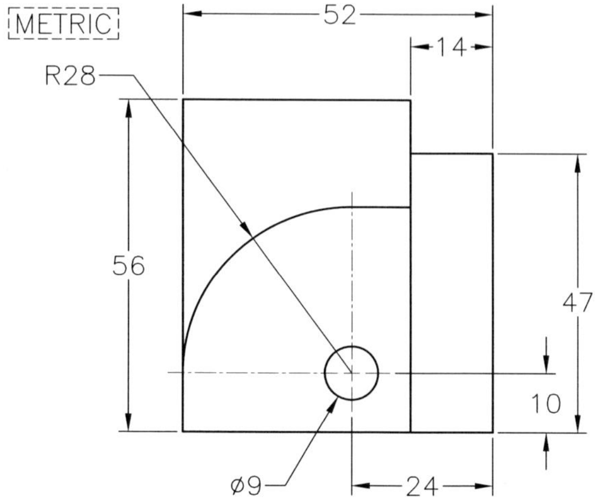

Figure 6-6 *Drawing for Exercise 2*

Chapter 7

Editing Dimensions

Learning Objectives

After completing this chapter, you will be able to:

- *Edit dimensions*
- *Stretch, extend, and trim dimensions*
- *Use the Oblique and Text Angle command options to edit dimensions*
- *Update dimensions using the Update command*
- *Use the PROPERTIES palette to edit dimensions*

EDITING DIMENSIONS USING EDITING TOOLS

For editing dimensions, AutoCAD has provided some special editing tools that work with dimensions. These editing tools can be used to define a new dimension text, return to the home text, create oblique dimensions, and rotate and update the dimension text. You can also use the **Trim**, **Stretch**, and **Extend** tools to edit the dimensions. The properties of the dimensioned objects can also be changed using the **PROPERTIES** palette or the **Dimension Style Manager**.

Editing Dimensions by Stretching

You can edit a dimension by stretching it. However, to stretch a dimension, appropriate definition points must be selected by using the crossing or window method. As the middle point of the dimension text is a definition point for all types of dimensions, you can easily stretch and move the dimension text to any location you want.

Tutorial 1 — Edit by Stretching

In this tutorial, you will stretch the objects and dimensions shown in Figure 7-1 using grips. The new location of the lines and dimensions is at a distance of 0.5 units in the positive *Y* axis direction, as shown in Figure 7-2.

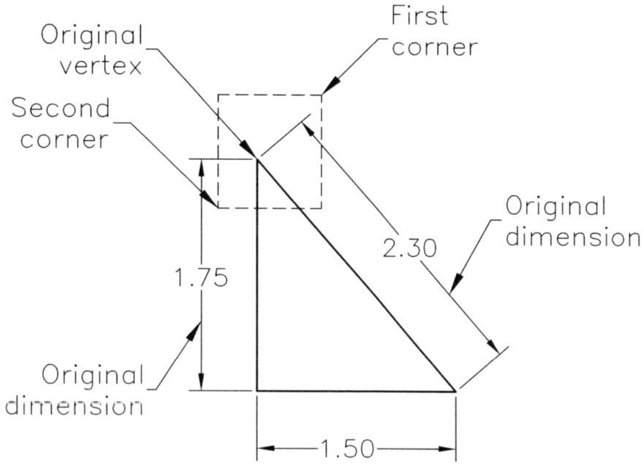

Figure 7-1 *Original location of lines and dimensions*

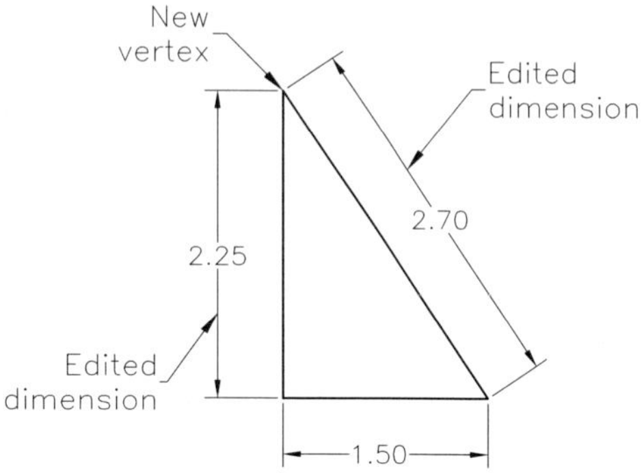

Figure 7-2 *New location of lines and dimensions*

1. Choose the **Stretch** tool from the **Modify** panel of the **Home** tab. The prompt sequence after choosing this tool is given below:

   Select objects to stretch by crossing-window or crossing-polygon
   Select objects: *Define a crossing window using the first and second corners, as shown in Figure 7-1.*
   Select objects: Enter
   Specify base point or [Displacement]<Displacement>: *Select original vertex using the osnap as the base point.*
   Specify second point or <use first point as displacement>: **@0.5<90**

2. The selected entities will be stretched to the new location. The dimension that was initially 1.75 will become 2.25 and the dimension that was initially 2.30 will become 2.70, refer to Figure 7-2.

Editing Dimensions by Trimming and Extending

Trimming and extending operations can be carried out with all types of linear dimensions (horizontal, vertical, aligned, and rotated) and the ordinate dimension. Even if the dimensions are true associative, you can extend and trim them, refer to Figures 7-3 and 7-4. AutoCAD trims or extends a linear dimension between the extension line definition points and the object used as a boundary or trimming edge. To extend or trim an ordinate dimension, AutoCAD moves the feature location (location of the dimensioned coordinate) to the boundary edge. To retain the original ordinate value, the boundary edge to which the feature location point is moved should be orthogonal to the measured ordinate. In both cases, the imaginary line drawn between the two extension line definition points is trimmed or extended by AutoCAD, and the dimension is adjusted automatically.

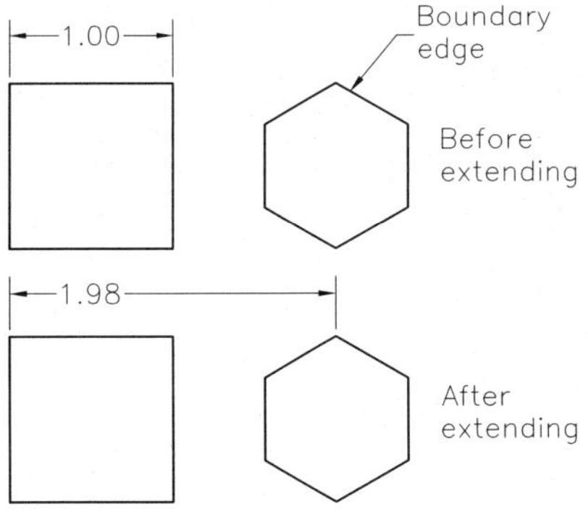

Figure 7-3 *Dimensions edited by extending*

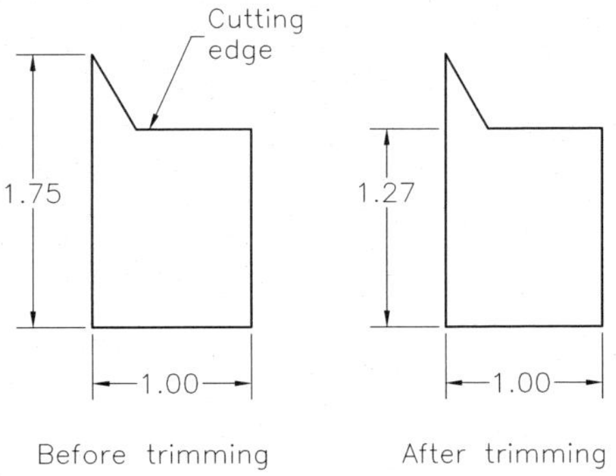

Figure 7-4 *Edgemode extended trimming*

Flipping Dimension Arrow

You can flip the arrowheads individually. To flip the arrow, select the dimension. Place the cursor on the grip corresponding to the arrowhead that you want to flip. When the color of the grip turns pink, a shortcut menu will be invoked. Now, choose the **Flip Arrow** option from the shortcut menu.

MODIFYING THE DIMENSIONS

Toolbar: Dimension > Dimension Edit **Command:** DIMEDIT

The dimensions can be modified by choosing the **Dimension Edit** tool from the **Dimension** toolbar, refer to Figure 7-5. Alternatively, you can use the **DIMEDIT** command to modify the dimensions. This command has four options: **Home**, **New**, **Rotate**, and **Oblique**. The prompt sequence that will follow when you choose this tool is given below:

Enter type of dimension editing [Home/New/Rotate/Oblique] <Home>: *Enter an option.*

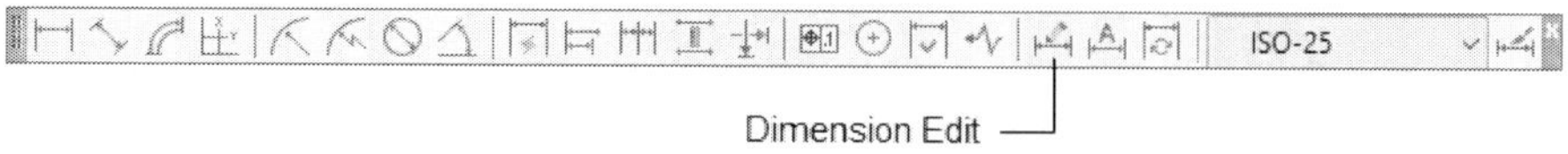

***Figure 7-5** Choosing the **Dimension Edit** tool from the **Dimension** toolbar*

Home

The **Home** option restores the text of a dimension to its original (home/default) location if the position of the text has been changed by stretching or editing, refer to Figure 7-6.

New

The **New** option is used to replace the existing dimension with a new text string. When you invoke this option, the **Text Editor** will be displayed. By default, 0.0000 will be displayed. Using the **Text Editor**, enter the dimension or write the text string with which you want to replace the existing dimension. Once you have entered a new dimension in the editor and exit the **Text Editor**, you will be prompted to select the dimension to be replaced. Select the dimension and press ENTER; it will be replaced with the new dimension, refer to Figure 7-6.

Rotate

The **Rotate** option is used to position the dimension text at the specified angle. With this option, you can change the orientation (angle) of the dimension text of any number of associative dimensions. The angle can be specified by entering a value at the **Specify angle for dimension text** prompt or by specifying two points at the required angle. Once you have specified the angle, you will be prompted to select the dimension text to be rotated. Select the dimension and press ENTER; the text will rotate about its middle point, refer to Figure 7-6. You can also invoke this option by choosing the **Text Angle** tool from the extended **Dimensions** panel of the **Annotate** tab.

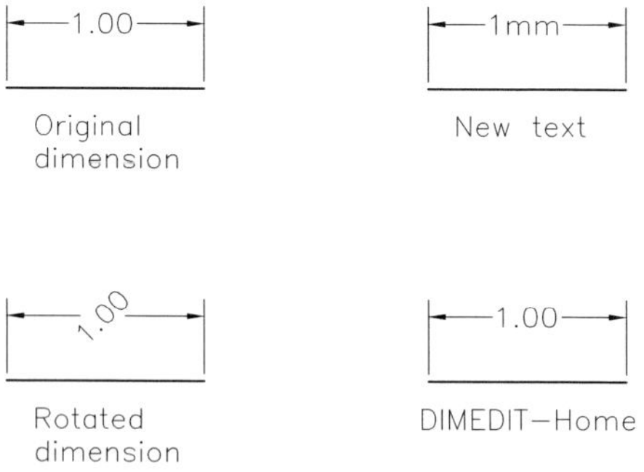

***Figure 7-6** Using the **Dimension Edit** tool to edit dimensions*

Oblique

In linear dimensions, extension lines are drawn perpendicular to the dimension line. The **Oblique** option bends the linear dimensions. It draws extension lines at an oblique angle, refer to Figure 7-7.

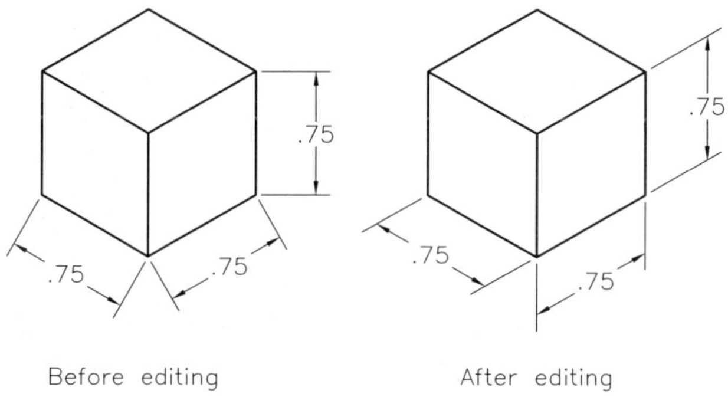

Figure 7-7 *Using the* ***Oblique*** *option to edit dimensions*

EDITING THE DIMENSION TEXT

Toolbar: Dimension > Dimension Text Edit **Command:** DIMTED or DIMTEDIT
Menu Bar: Dimension > Align Text

The dimension text can be edited by using the **Dimension Text Edit** tool from the **Dimension** toolbar. This tool is used to edit the placement and orientation of a single existing dimension. You can use this tool in cases where dimension texts of two or more dimensions are too close together. In such cases, the **Dimension Text Edit** tool is invoked to move the dimension text to some other location so that there is no confusion.

UPDATING DIMENSIONS

Ribbon: Annotate > Dimensions > Update **Toolbar:** Dimension > Dimension Update
Menu Bar: Dimension > Update

The **Update** tool is used to regenerate and update the prevailing dimension entities (such as arrows heads and text height) using the current settings for the dimension variables, dimension style, text style, and units. On choosing this tool, you will be prompted to select the dimensions to be updated. You can select all the dimensions or specify those that should be updated.

EDITING DIMENSIONS USING THE PROPERTIES PALETTE

Ribbon: View > Palettes > Properties **Toolbar:** Standard > Properties
Menu Bar: Modify > Properties **Command:** PROPERTIES/PR/CH/MO

You can also modify a dimension or leader by using the **PROPERTIES** palette. The **PROPERTIES** palette is displayed when you choose the **Properties** button from the **Palettes** panel of the **View** tab. Alternatively, you can invoke the **PROPERTIES** palette by choosing the **Properties** option from the **Palettes** flyout from the **Tools** menu. All the **properties** of the selected object are displayed in the **PROPERTIES** palette. Select the dimension before invoking the **PROPERTIES** palette, otherwise it would not give the description of the dimension.

Tutorial 2 Modify Dimensions

In this tutorial, you will modify the dimensions given in Figure 7-8 so that they match the dimensions given in Figure 7-9.

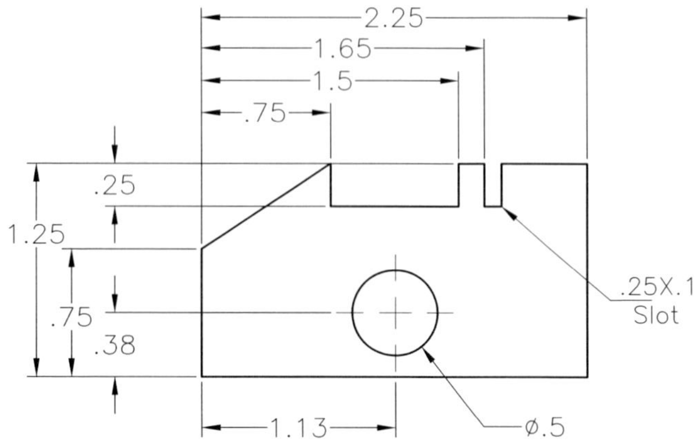

Figure 7-8 Drawing for Tutorial 2

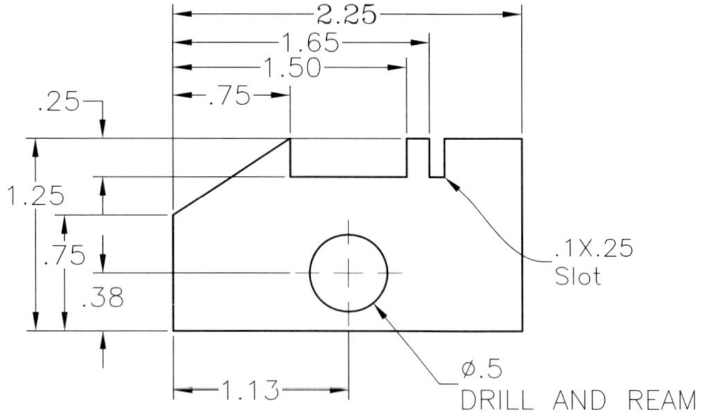

Figure 7-9 Drawing after editing the dimensions

1. Choose the **Text Style** (inclined arrow) tool from the **Text** panel of the **Annotate** tab and create a new text style with the name **ROMANC**. Select **romanc.shx** as the font for the style.

2. Select the dimension **2.25** and enter **PR** at the command prompt to display the **PROPERTIES** palette.

3. In the **Text** category, select the **Text style** drop-down list and then select the **ROMANC** style from this drop-down list; the changes will take place dynamically.

4. Once all the required changes are made in the linear dimension, choose the **Select Objects** button in the **PROPERTIES** palette; you will be prompted to select the object. Select the leader line and then press ENTER.

5. The **PROPERTIES** palette will display the leader options. Select **Spline** from the **Leader type** drop-down list in the **Leaders** category; the straight line will be dynamically converted into a spline with an arrow.

6. Close the **PROPERTIES** palette.

7. Choose the **Dimension Edit** tool from the **Dimension** toolbar. Enter **N** in the prompt sequence to display the **Text Editor**.

8. Enter **%%C.5 DRILL AND REAM** in the **Text Editor** and then click outside it to accept the changes. You will be prompted to select the object to be changed.

9. Select the diameter dimension and then press ENTER; the diameter dimension will be modified to the new value.

EXERCISES

Exercise 1 — Edit Dimensions

1. Create the drawing shown in Figure 7-10.
2. Dimension the drawing, as shown in Figure 7-10.
3. Edit the dimensions so that they match the dimensions shown in Figure 7-11.

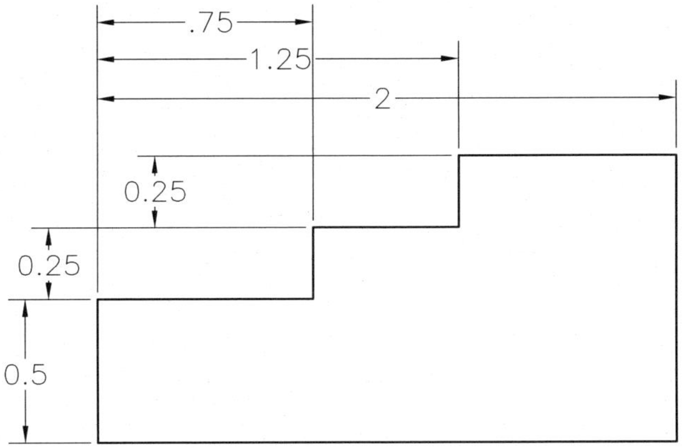

Figure 7-10 *Drawing for Exercise 1 before editing dimensions*

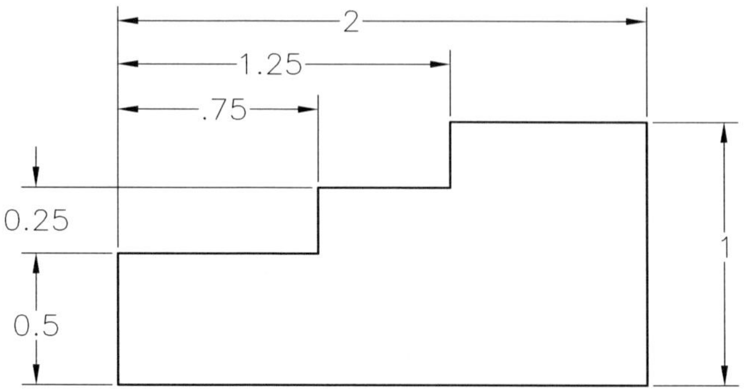

Figure 7-11 *Drawing for Exercise 1 after editing dimensions*

Exercise 2 Edit Dimensions

1. Draw the object shown in Figure 7-12(a). Assume the dimensions where necessary.
2. Dimension the drawing, as shown in Figure 7-12(a).
3. Edit the dimensions so that they match the dimensions shown in Figure 7-12(b).

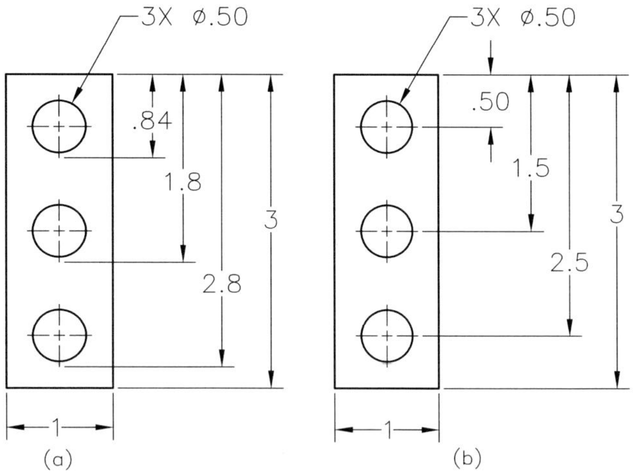

Figure 7-12 *Drawings for Exercise 2*

Chapter 8

Dimension Styles, Multileader Styles, and System Variables

Learning Objectives

After completing this chapter, you will be able to:

- *Use styles and variables to control dimensions*
- *Create dimension styles*
- *Set dimension variables*
- *Use dimension style overrides*
- *Compare and list dimension styles*
- *Import externally referenced dimension styles*
- *Create, restore, and modify multileader styles*

USING STYLES AND VARIABLES TO CONTROL DIMENSIONS

In AutoCAD, the appearance of dimensions in the drawing area and the manner in which the dimensions are saved in the drawing database are controlled by a set of dimension variables. The dimensioning tools use these variables as arguments. The variables that control the appearance of dimensions can be managed using dimension styles. You can use the **Dimension Style Manager** dialog box to control the dimension styles and dimension variables through a set of dialog boxes.

CREATING AND RESTORING DIMENSION STYLES

Ribbon: Annotate > Dimensions > Dimension Style (The inclined arrow)
Toolbar: Dimension > Dimension Style or Styles > Dimension Style
Command: DIMSTYLE/D

The dimension styles control the appearance and positioning of dimensions and leaders in the drawing. If the default dimensioning style (Standard, Annotative or ISO 25) does not meet your requirements, you can select another existing dimensioning style or create a new one. The default names of the dimension style files are **Standard**, **ISO 25**, and **Annotative**. Dimension styles can be created by using the **Dimension Style Manager** dialog box. Left-click on the inclined arrow in the **Dimensions** panel of the **Annotate** tab to invoke the **Dimension Style Manager** dialog box. Figure 8-1 shows the **Dimension Style Manager** dialog box with the **Standard** dimension style chosen in the **Styles** list box.

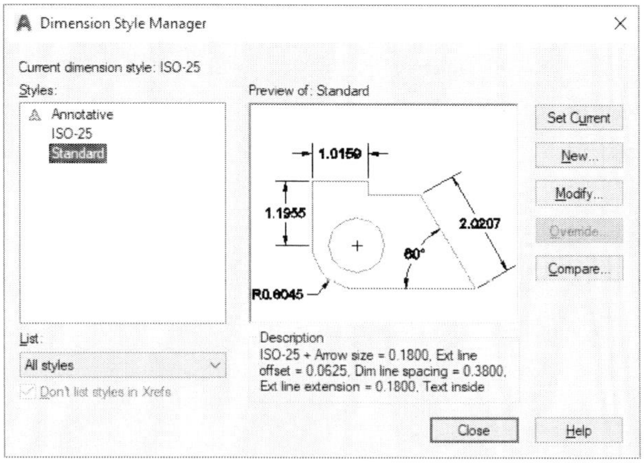

Figure 8-1 *The* ***Dimension Style Manager*** *dialog box*

In the **Dimension Style Manager** dialog box, choose the **New** button to display the **Create New Dimension Style** dialog box.

NEW DIMENSION STYLE DIALOG BOX

The **New Dimension Style** dialog box can be used to specify the dimensioning attributes (variables) that affect various properties of the dimensions. The various tabs provided under the **New Dimension Style** dialog box are discussed next.

Lines Tab

The options in the **Lines** tab of the **New Dimension Style** dialog box are used to specify the dimensioning attributes (variables) that affect the format of the dimension lines. For example, the appearance and behavior of the dimension lines and extension lines can be changed in this tab.

Symbols and Arrows Tab

The options in the **Symbols and Arrows** tab of the **New Dimension Style** dialog box are used to specify the variables and attributes that affect the format of the symbols and arrows. You can change the appearance of symbols and arrows.

CONTROLLING THE DIMENSION TEXT FORMAT

Text Tab

You can control the dimension text format through the **Text** tab of the **New Dimension Style** dialog box. In the **Text** tab, you can control the parameters such as the placement, appearance, horizontal and vertical alignment of the dimension text, and so on. For example, you can force AutoCAD to align the dimension text along the dimension line. You can also force the dimension text to be displayed at the top of the dimension line. You can save the settings in a dimension style file for future use. The **New Dimension Style** dialog box has the **Preview** window that updates dynamically to display the text placement as the settings are changed. Individual items of the **Text** tab and the related dimension variables are described next.

FITTING DIMENSION TEXT AND ARROWHEADS

Fit Tab

The **Fit** tab provides you with the options that are used to control the placement of dimension lines, arrowheads, leader lines, text, and the overall dimension scale.

FORMATTING PRIMARY DIMENSION UNITS

Primary Units Tab

You can use the **Primary Units** tab of the **New Dimension Style** dialog box to control the dimension text format and precision values. You can use the options under this tab to control Unit format, Precision, and Zero suppression for dimension measurements. AutoCAD lets you attach a user-defined prefix or suffix to the dimension text. For example, you can define the diameter symbol as a prefix by entering %%C in the **Prefix** edit box; AutoCAD will automatically attach the diameter symbol in front of the dimension text. Similarly, you can define a unit type, such as **mm**, as a suffix; AutoCAD will then attach **mm** at the end of every dimension text. This tab also enables you to define zero suppression, precision, and dimension text format.

FORMATTING ALTERNATE DIMENSION UNITS

Alternate Units Tab

By default, the options in the **Alternate Units** tab of the **New Dimension Style** dialog box are disabled. If you want to perform alternate units dimensioning, select the **Display alternate units** check box. By doing so, AutoCAD activates various options in this area. This tab sets the format, precision, angles, placement, scale, and so on for the alternate units in use. In this tab, you can specify the values that will be applied to the alternate dimensions.

FORMATTING THE TOLERANCES

Tolerances Tab

The **Tolerances** tab allows you to set the parameters for options that control the format and display of the tolerance dimension text. These include the alternate unit tolerance dimension text.

DIMENSION STYLE FAMILIES

The dimension style feature of AutoCAD lets the user define a dimension style with values that are common to all dimensions. For example, the arrow size, dimension text height, and color of the dimension line are generally same in all types of dimensioning such as linear, radial, diameter, and angular. These dimensioning types belong to the same family because they have some characteristics in common. In AutoCAD, this is called a dimension style family, and the values assigned to the family are called dimension style family values.

After you have defined the dimension style family values, you can specify variations on it for other types of dimensions such as radial and diameter. For example, if you want to limit the number of decimal places to two in radial dimensioning, you can specify that value for radial dimensioning. The other values will stay the same as the family values to which this dimension type belongs. When you use the radial dimension, AutoCAD automatically uses the style that was defined for radial dimensioning; otherwise, it creates a radial dimension with the values as defined for the family. After you have created a dimension style family, any changes in the parent style are applied to family members, if the particular property is the same. Special suffix codes are appended to the dimension style family names that correspond to different dimension types. For example, if the dimension style family name is MYSTYLE and you define a diameter type of dimension, AutoCAD will append $4 at the end of the dimension style family name. The name of the diameter type of dimension will be MYSTYLE$4. The following are the suffix codes for different types of dimensioning.

Suffix Code	Dimension Type	Suffix Code	Dimension Type
0	Linear	2	Angular
3	Radius	4	Diameter
6	Ordinate	7	Leader

Tutorial 1 Dimension Style Family

The following tutorial illustrates the concepts of dimension style families, refer to Figure 8-2.

1. Specify the values for the dimension style family.
2. Specify the values for the linear dimensions.
3. Specify the values for the diameter and radius dimensions.
4. After creating the dimension style, use it to dimension the given drawing.

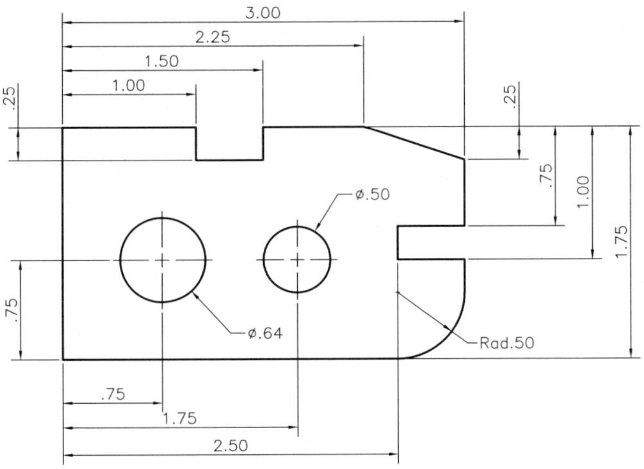

Figure 8-2 *Drawing for Tutorial 1*

The steps for creating a dimension style family are given next:

1. Start a new file in the **Drafting & Annotation** workspace and draw an object, as shown in Figure 8-2.

2. Left-click on the inclined arrow in the **Dimensions** panel of the **Annotate** tab; the **Dimension Style Manager** dialog box is displayed.

3. Choose the **New** button to display the **Create New Dimension Style** dialog box. In this dialog box, enter **MyStyle** in the **New Style Name** edit box. Select **Annotative** from the **Start With** drop-down list. Also, select **All dimensions** from the **Use for** drop-down list.

   Now, choose the **Continue** button to display the **New Dimension Style: MyStyle** dialog box. In this dialog box, Choose the **Lines** and **Symbols and Arrows** tab and then enter the following values in their respective options.

   Lines tab
   Baseline Spacing: 0.15
   Offset from origin: 0.03
   Extend beyond dim lines: 0.07

   Symbols and Arrows tab
   Arrow size: 0.09
   Center marks: Select the **Line** radio button
   Enter **0.05** in the spinner in the **Center marks** area

4. Choose the **Text** tab and change the following values:

   Text Height: 0.09
   Offset from dim line: 0.03

5. Choose the **Fit** tab and make sure the **Annotative** check box is selected.

6. After entering the values, choose the **OK** button to return to the **Dimension Style Manager** dialog box. This dimension style contains the values that are common to all dimension types.

7. Now, choose the **New** button again in the **Dimension Style Manager** dialog box to display the **Create New Dimension Style** dialog box. AutoCAD displays **Copy of MyStyle** in the **New Style Name** edit box. Select **MyStyle** from the **Start With** drop-down list if it is not already selected. From the **Use for** drop-down list, select **Linear dimensions**. Choose the **Continue** button to display the **New Dimension Style: MyStyle: Linear** dialog box and set the following values in the **Text** tab:

    a. Select the **Aligned with dimension line** radio button in the **Text alignment** area.
    b. In the **Text placement** area, select **Above** from the **Vertical** drop-down list.

8. In the **Primary Units** tab, set the precision to two decimal places by selecting **0.00** from the **Precision** drop-down list. Select the **Leading** check box in the **Zero suppression** area. Next, choose the **OK** button to return to the **Dimension Style Manager** dialog box.

9. Choose the **New** button again to display the **Create New Dimension Style** dialog box. Select **MyStyle** from the **Start With** drop-down list and the **Diameter dimensions** type from the **Use for** drop-down list. Next, choose the **Continue** button; the **New dimension Style: MyStyle: Diameter** dialog box is displayed.

10. Choose the **Text** tab and select the **Centered** option from the **Vertical** drop-down list in the **Text placement** area, if it is not selected by default. Then select the **Horizontal** radio button from the **Text alignment** area, if it is not selected by default.

11. Choose the **Primary Units** tab and then set the precision to two decimal places. Select the **Leading** check box in the **Zero suppression** area. Next, choose the **OK** button to return to the **Dimension Style Manager** dialog box.

12. In this dialog box, again choose the **New** button to display the **Create New Dimension Style** dialog box. Select **MyStyle** from the **Start With** drop-down list and **Radius dimensions** from the **Use for** drop-down list. Choose the **Continue** button to display the **New Dimension Style: MyStyle: Radial** dialog box.

13. Choose the **Primary Units** tab and then set the precision to two decimal places. Select the **Leading** check box in the **Zero suppression** area. Next, enter **Rad** in the **Prefix** edit box.

14. In the **Fit** tab, select the **Text** radio button in the **Fit options** area. Next, choose the **OK** button to return to the **Dimension Style Manager** dialog box.

15. Select **MyStyle** from the **Styles** list box in the **Dimension Style Manager** dialog box and choose the **Set Current** button. Choose the **Close** button to exit the dialog box.

16. Choose the **Linear** tool from the **Dimensions** panel; the **Select Annotation Scale** dialog box is displayed. Select the **1:1** option from the drop-down list and choose the **OK** button.

17. Use the linear and baseline dimensions to draw the linear dimensions, refer to Figure 8-2. While entering linear dimensions, you will notice that AutoCAD automatically uses the values that were defined for the linear type of dimensioning.

18. Use the diameter dimensioning to dimension the circles, refer to Figure 8-2. Again, notice that the dimensions are drawn based on the values specified for the diameter type of dimensioning.

19. Now, use the radius dimensioning to dimension the fillet, refer to Figure 8-2.

USING DIMENSION STYLE OVERRIDES

Most of the dimension characteristics are common in a production drawing. The values that are common to different dimensioning types can be defined in the dimension style family. However, at times, you might have different dimensions. For example, you may need two types of linear dimensioning: one with tolerance and one without it. One way to draw these dimensions is to create two dimensioning styles. You can also use the dimension variable overrides to override the existing values. For example, you can define a dimension style (**MyStyle**) that draws dimensions without tolerance. Now, to draw a dimension with tolerance or update an existing dimension, you can override the previously defined value through the **Dimension Style Manager** dialog box or by setting the variable values at the command prompt. Now, you can remove style override of any selected dimension by choosing the **Remove Style Overrides** option from the shortcut menu displayed on right clicking. The following tutorial illustrates how to use the dimension style overrides.

Tutorial 2 — Dimension Style Override

In this tutorial, you will update the overall dimension (3.00) so that the tolerance is displayed with the dimension. You will also add linear dimensions, as shown in Figure 8-3.

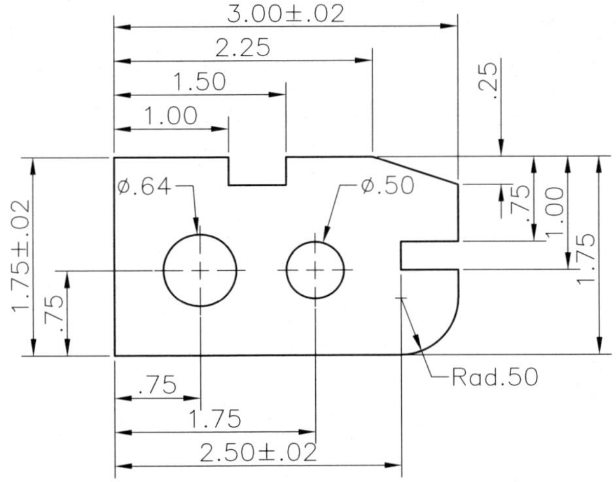

Figure 8-3 *Drawing for Tutorial 2*

This problem can be solved by using dimension style overrides as well as by using the **PROPERTIES** palette. However, here only the dimension style overrides method is discussed.

1. Invoke the **Dimension Style Manager** dialog box. Select **MyStyle** from the **Styles** list box and choose the **Override** button to display the **Override Current Style: MyStyle** dialog box.

The options in this dialog box are same to the **New Dimension Style** dialog box discussed earlier in this chapter.

2. Choose the **Tolerances** tab and select **Symmetrical** from the **Method** drop-down list.

3. Set the value of the **Precision** spinner to two decimal places. Set the value of the **Upper value** spinner to **0.02** and select the **Leading** check box in the **Zero suppression** area. Next, choose the **OK** button to exit the dialog box (this does not save the style). On doing so, you will notice that **<style overrides>** is displayed under **MyStyle** in the **Styles** list box, indicating that the style overrides the **MyStyle** dimension style.

4. The **<style overrides>** is displayed until you save it under a new name or under the style it is displayed in, or until you delete it. Select **<style overrides>** and right-click to display the shortcut menu. Choose the **Save to current style** option from the shortcut menu to save the overrides to the current style. Choosing the **Rename** option allows you to rename the style override and save it as a new style.

5. Choose the **Update** tool from the **Dimension** panel in the **Annotate** tab and select the dimension that measures **3.00**. The dimension now displays the symmetrical tolerance.

6. Draw the remaining two linear dimensions. They will automatically appear with the tolerances, refer to Figure 8-3.

COMPARING AND LISTING DIMENSION STYLES

Choosing the **Compare** button in the **Dimension Style Manager** dialog box displays the **Compare Dimension Styles** dialog box where you can compare the settings of two dimensions styles or list all the settings of a single dimension style.

USING EXTERNALLY REFERENCED DIMENSION STYLES

The externally referenced dimensions cannot be used directly in the current drawing. When you Xref a drawing, the drawing name is appended to the style name and the two are separated by a vertical bar (|) symbol. It uses the same syntax as the other externally dependent symbols. For example, if the drawing (FLOOR) has a dimension style called DIM1 and you Xref this drawing in the current drawing, AutoCAD will rename the dimension style to FLOOR|DIM1. You cannot make this dimension style current, nor can you modify or override it. However, you can use it as a template to create a new style.

CREATING AND RESTORING MULTILEADER STYLES

Ribbon: Annotate > Leaders > Multileader Style Manager (Inclined arrow)
Toolbar: Multileader > Multileader Style or Styles > Multileader Style
Command: MLEADERSTYLE/MLS

The multileader styles control the appearance and positioning of multileaders in the drawing. If the default multileader styles (**Standard** and **Annotative**) do not meet your requirement, you can edit the default multileader styles or create a new one as per your requirement.

MODIFY MULTILEADER STYLE DIALOG BOX

The **Modify Multileader Style** dialog box can be used to specify the multileader attributes (variables) that affect various properties of the multileader. The various tabs provided in the **Modify Multileader Style** dialog box are discussed next.

Leader Format Tab

The options in the **Leader Format** tab of the **Modify Multileader Style** dialog box are used to specify the multileader attributes that affect the format of the multileader lines.

Leader Structure Tab

The options in the **Leader Structure** tab of the **Modify Multileader Style** dialog box are used to specify the dimensioning attributes that affect the structure of the multileader lines. The attributes that can be controlled by using the **Leader Structure** tab are the number of lines to be drawn before adding the content, adding landing before the content, length of the landing line, multileader to be annotative or not, and so on.

Content Tab

The options in the **Content** tab of the **Modify Multileader Style** dialog box are used to specify the multileader attributes that affect the content and the format of the text or block to be attached with the multileader.

EXERCISES

Exercises 1 and 2

Create the drawings shown in Figures 8-4 and 8-5. You must create dimension style and multileader style files and specify the values for the different dimension types such as linear, radial, diameter, and ordinate.

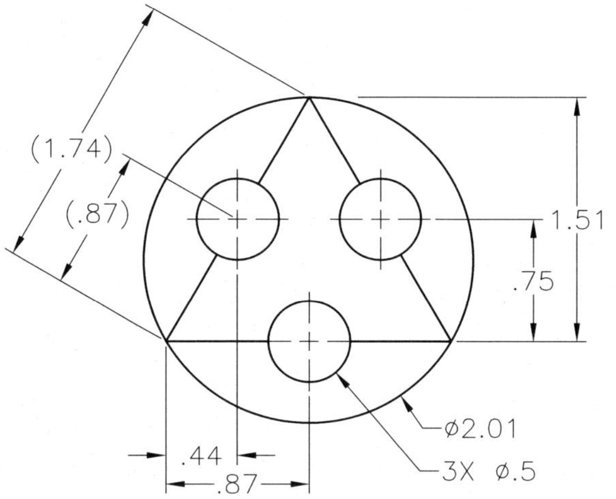

Figure 8-4 *Drawing for Exercise 1*

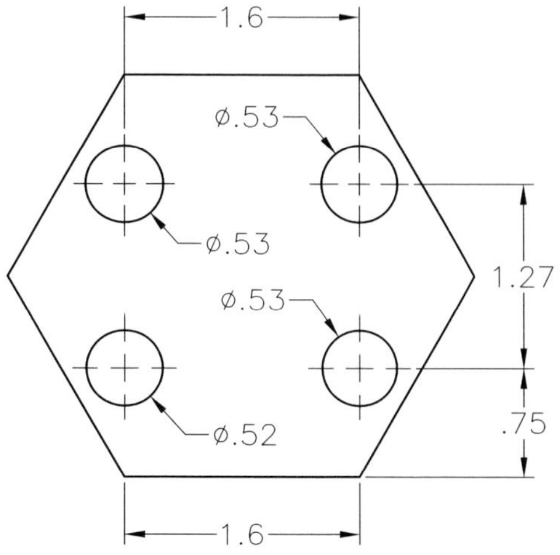

Figure 8-5 *Drawing for Exercise 2*

Exercise 3 — Dimension Style

Draw the sketch shown in Figure 8-6. You must create the dimension style and specify different dimensioning parameters in the dimension style.

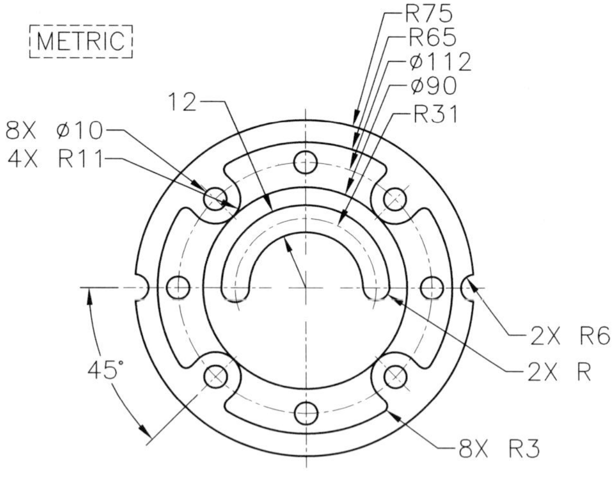

Figure 8-6 *Drawing for Exercise 3*

Chapter 9

Creating Texts and Tables

Learning Objectives

After completing this chapter, you will be able to:

- *Understand the use of annotative objects*
- *Write text using the Single Line and Multi Line Text tools*

ANNOTATIVE OBJECTS

AutoCAD provides improved functionality of the drawing annotation that enables you to annotate your drawing easily and efficiently. Annotations are the notes and objects that are inserted into a drawing to add important information to it. The list of annotative objects is given next.

Text	Mtext	Dimensions	Leaders	Blocks
Multileaders	Hatches	Tolerances	Attributes	

ANNOTATION SCALE

Annotation Scale is a feature that allows you to control the size of the annotative objects in model space.

Generally, the formula used to calculate the height of the annotative objects in model space is:

Height of the annotative object in model space = Annotation scale X Height of the annotative object in paper space.

MULTIPLE ANNOTATION SCALES

Ribbon: Annotate > Annotation Scaling > Add/Delete Scales **Command:** OBJECTSCALE
Menu Bar: Modify > Annotative Object Scale > Add/Delete Scales

You can assign multiple annotation scales to an annotative object. This enables you to display or print the same annotative object in different sizes. The annotative object is displayed or printed based on the current annotation scale. Assigning multiple annotation scales to annotative objects saves a considerable amount of time that is lost while creating a set of objects with different scales in different layers. You can add multiple annotation scales to objects manually and automatically.

CONTROLLING THE DISPLAY OF ANNOTATIVE OBJECTS

After assigning annotation scale to objects, the paper height display scale (in case of plotting) or the viewport scale (in case of viewports) is automatically applied to annotative objects and they are displayed accordingly in the model space or viewport. The annotative objects that do not support the current annotation scale will not be scaled and will be displayed according to the previous annotation scale applied to them.

You can control the display of annotative objects by using the **Show annotation objects** button in the Status Bar. If this button is turned on, all annotative objects will be displayed in the model space or viewport regardless of the current annotation scale. If this button is turned off, only the annotative objects that have annotation scale equal to that of the current annotation scale will be displayed. Note that the **Show annotation objects** buttons in the model and layout tabs are independent of each other.

CREATING TEXT

In manual drafting, lettering is accomplished by hand using a lettering device, pen, or pencil. This is a very time-consuming and tedious job. Computer-aided drafting has made this process extremely easy. Engineering drawings invoke certain standards to be followed in connection with the placement of a text in a drawing. In this section, you will learn how a text can be added in a drawing by using the **Single Line** and **Multiline Text** tools.

Writing Single Line Text

Ribbon: Home > Annotation > Text drop-down > Single Line Or
Annotate > Text > Text drop-down > Single Line **Toolbar:** Text > Single Line Text
Menu Bar: Draw > Text > Single Line Text **Command:** TEXT/DTEXT/DT

The **Single Line** tool is used to write text on a drawing. While writing, you can delete the typed text by using the BACKSPACE key. On invoking the **Single Line** tool, you will be prompted to specify the start point. The default and the most commonly used option in the **Single Line** tool is the **start point** option. By specifying a start point, the text is left-justified along its baseline. Baseline refers to the line along which its base lies. After specifying the start point, you need to set the height and the rotation angle of the text.

The **Specify height <current>:** prompt determines the distance by which the text extends above the baseline, measured by the capital letters. This distance is specified in drawing units. You can specify the text height by specifying two points or by entering a value. In the case of a null response, the default height, that is, the height used for the previous text drawn in the same style will be used.

The **Specify rotation angle of text <current>:** prompt determines the angle at which the text line will be drawn. The default value of the rotation angle is 0-degree (along east); and in this case, the text is drawn horizontally from the specified start point. The rotation angle is measured in counterclockwise direction. The last angle specified becomes the current rotation angle and if you give a null response, the last angle specified will be used as default rotation angle. You can also specify the rotation angle by specifying two points. The text will be drawn upside down if a point is specified at a location to the left of the start point.

You can now enter the text in the text window. The characters will be displayed as you type them. After entering the text, click outside the text window and press ESC. The prompt sequence displayed on choosing this tool is given next.

Specify start point of text or [Justify/Style]: *Specify the start point.*
Specify height <0.2000>: **0.15** Enter
Specify rotation angle of text <0>: Enter; *the textbox will be displayed. Start typing in it.*

ENTERING SPECIAL CHARACTERS

In almost all drafting applications, you need to use special characters (symbols) in the normal text and in the dimension text. For example, you may want to use the degree symbol (°) or the diameter symbol (ø), or you may want to underscore or overscore some text. This can be achieved with the appropriate sequence of control characters (control code). For each symbol, the control sequence starts with a percent sign written twice (%%). The character immediately following the double percent sign depicts the symbol. The control sequences for some of the symbols are given next.

Control sequence	Special character
%%c	Diameter symbol (ø)
%%d	Degree symbol (°)
%%p	Plus/minus tolerance symbol (±)

For example, if you want to write 25° Celsius, you need to enter **25%%dCelsius**. If you want to write 43.0ø, you need to enter **43.0%%c**.

CREATING MULTILINE TEXT

Ribbon: Home > Annotation > Text drop-down > Multiline Text **Command:** MTEXT
Menu Bar: Draw > Text > Multiline Text **Tool Palette:** Draw > MText
Toolbar: Draw> Multiline Text Or Text > Multiline Text

The **Multiline Text** tool in the **Text** drop-down of the **Text** panel is used to write a multiline text whose width is specified by defining two corners of the text boundary or by entering a width using the coordinate entry. The text created by using the **Multiline Text** tool is a single object regardless of the number of lines it contains. On invoking the **Multiline Text** tool, a sample text "abc" is attached to the cursor and you are prompted to specify the first corner. Specify the first corner and move the pointing device so that a box that shows the location and size of the paragraph text is formed. An arrow is displayed within the boundary indicating the direction of the text flow. Specify the other corner to define the boundary. When you define the text boundary, it does not mean that the text paragraph will fit within the defined boundary. AutoCAD only uses the width of the defined boundary as the width of the text paragraph. The height of the text boundary has no effect on the text paragraph.

The prompt sequence that will be displayed after choosing the **Multiline Text** tool is given next.

Specify first corner: *Select a point to specify first corner.*
Specify opposite corner or [Height/Justify/Line spacing/Rotation/Style/Width/Columns]: *Select an option or select a point to specify other corner. Now, you can write text in this boundary.*

Tutorial 1 — *Multiline Text*

In this tutorial, you will use the **Multiline Text** tool to write the following text.

For long, complex entries, create multiline text using the MTEXT option. The angle is 10°, dia = 1/2", and length = 32 1/2".

The font of the text is **Swis721 BT**, text height is 0.20, color is red, and written at an angle of 10-degree with Middle-Left justification. Make the word "multiline" bold, underline the text "multiline text", and make the word "angle" italic. The line spacing type and line spacing between the lines are **At least** and **1.5x**, respectively. Use the symbol for degrees. After writing the text in the text window, replace the word "option" with "command".

1. Choose the **Multiline Text** tool from the **Annotation** panel in the **Home** tab in the **Drafting & Annotation** workspace. After invoking this tool, specify the first corner on the screen to define the first corner of the paragraph text boundary. You need to specify the rotation

angle of the text before specifying the second corner of the paragraph text boundary. The prompt sequence is given next.

Current text style: "Standard" Text height: 0.2000 Annotative: No
Specify first corner: *Select a point to specify the first corner.*
Specify opposite corner or [Height/Justify/Line spacing/Rotation/Style/Width/Columns]: **R** Enter
Specify rotation angle <0>: 10 Enter
Specify opposite corner or [Height/Justify/Line spacing/Rotation/Style/Width/Columns]: **L** Enter
Enter line spacing type [At least/Exactly] <At least>: Enter
Enter line spacing factor or distance <1x>: **1.5x**
Specify opposite corner or [Height/Justify/Line spacing/Rotation/Style/Width/Columns]: *Select another point to specify the other corner.*

The text window and the **Text Editor** tab are displayed.

2. Select the **Swis721 BT** font from the **Font** drop-down list of the **Formatting** panel in the **Text Editor** tab.

3. Enter **0.20** in the **Text Height** edit box of the **Style** panel, if the value in this edit box is not 0.2.

4. Select **Red** from the **Color** drop-down list in the **Formatting** panel.

5. Now, enter the text in the text window, as shown in Figure 9-1. To add the degree symbol, choose the **Symbol** button from the **Insert** panel of the **Text Editor** tab in the **Ribbon**; a flyout is displayed. Choose **Degrees** from the flyout. When you type 1/2 after dia = and then press the " key, AutoCAD displays a small yellow icon. Click on the icon; a flyout will appear. Choose the **Diagonal** option from the flyout.

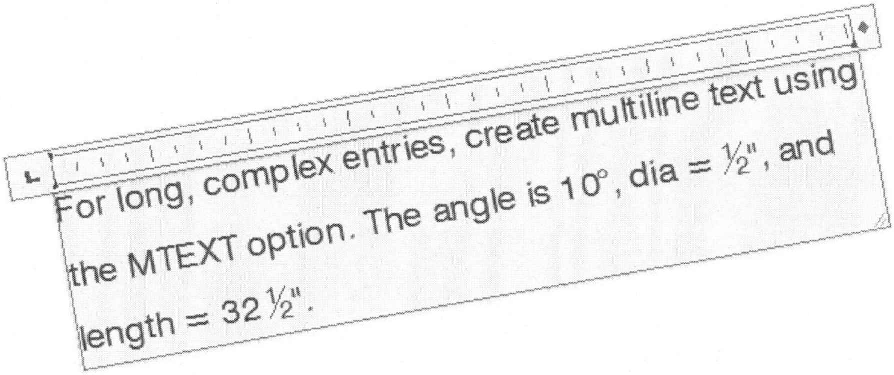

Figure 9-1 *Text entered in the text window*

Similarly, when you type 1/2 after length = 32 and then press the " key, AutoCAD displays a small yellow icon. Click on the icon; a flyout will appear. Choose the **Diagonal** option from the flyout.

6. Double-click on the word "multiline" to select it (or click and drag to select the text) and then choose the **B** button to make it boldface and underline the words "multiline text" by using the **U** button from the **Formatting** panel.

7. Highlight the word "angle" by double-clicking on it and then choose the **Italic** button from the **Formatting** panel.

8. Choose the **Middle Left ML** option from the **Justification** drop-down in the **Paragraph** panel.

9. Choose the **Find & Replace** button from the **Tools** panel in the **Text Editor** tab. Alternatively, right-click on the text window and choose **Find and Replace** from the shortcut menu. You can also use CTRL+R keys. On doing so, the **Find and Replace** dialog box is displayed.

10. In the **Find what** edit box, enter **option** and in the **Replace with** edit box, enter **command**.

11. Choose the **Find Next** button. AutoCAD finds the word "option" and highlights it. Choose the **Replace** button to replace **option** by **command**. The **AutoCAD** information box is displayed informing you that AutoCAD has finished searching for the word. Choose **OK** to close the information box.

Note

To set the width of the multiline text objects, hold and drag the arrowhead on the right side of the ruler. In this tutorial, the text has been accommodated into three lines.

12. Now, choose **Close** to exit the **Find and Replace** dialog box. To exit the editor mode, click outside the text window; a text is displayed on the screen, as shown in Figure 9-2.

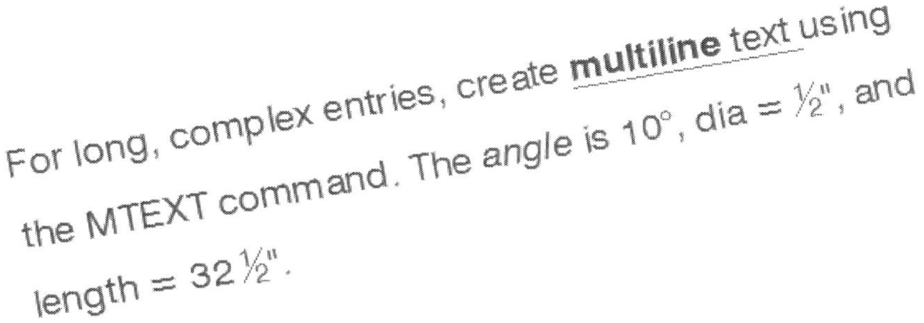

Figure 9-2 *Multiline text for Tutorial 1*

Tip

The text in the text window can be selected by double-clicking on the word or by triple-clicking on the text to select the entire line or paragraph.

EDITING TEXT

The contents of a Multiline Text object or a Single Line text object, or the text in a dimension object can be edited by using the **TEXTEDIT** command or the **PROPERTIES** palette. You can also use the AutoCAD editing tools such as **Move**, **Erase**, **Rotate**, **Copy**, and **Mirror** upon any text object.

EXERCISES

Exercise 1 *Single Line & Multiline Text*

Write the text using the **Single Line** and **Multiline Text** tools, as shown in Figure 9-3 and Figure 9-4. Use the special characters and the text justification options shown in the drawing. The text height is 0.1 and 0.15 respectively in Figure 9-3 and Figure 9-4.

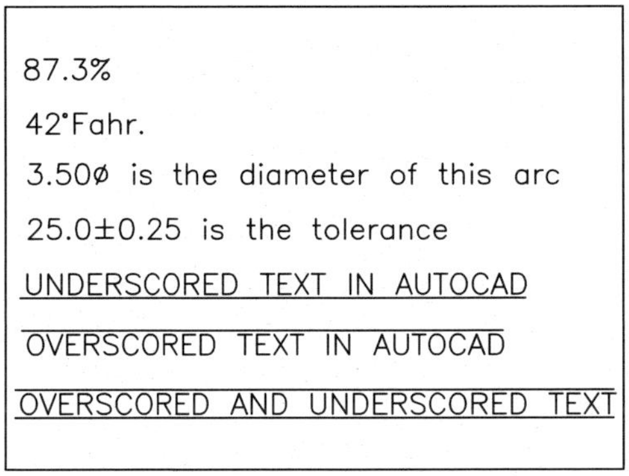

Figure 9-3 *Drawing for Exercise 1*

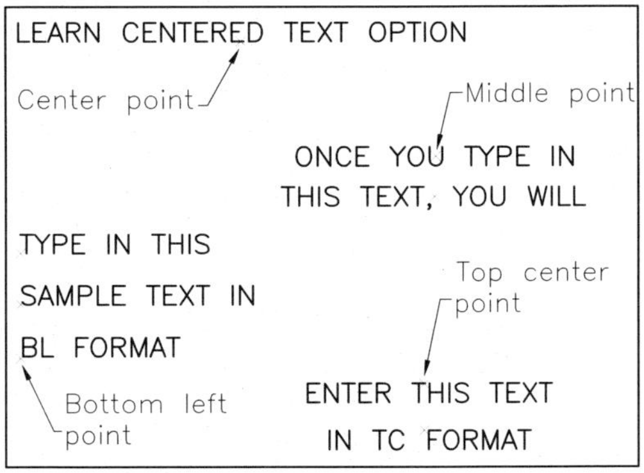

Figure 9-4 *Drawing for Exercise 1*

Exercise 2 *Text*

Write the text, shown in Figure 9-5, on the screen. Use the text justification that will produce the text as shown in the drawing. Assume a value for text height. Use the **PROPERTIES** palette to change the text, as shown in Figure 9-6.

Figure 9-5 *Drawing for Exercise 2*

Figure 9-6 *Drawing after changing the text*

Chapter 10

Editing Sketched Objects-II

Learning Objectives

After completing this chapter, you will be able to:

- *Use grips and adjust their settings*
- *Stretch, move, rotate, scale, and mirror objects with grips*
- *Use the Match Properties tool to match properties of the selected objects*
- *Use the Quick Select tool to select objects*
- *Manage contents using the DesignCenter*
- *Use the Inquiry tools*

INTRODUCTION TO GRIPS

Grips provide a convenient and quick means of editing objects. Grips are small squares that are displayed on the key points of an object on selecting the object, as shown in Figure 10-1.

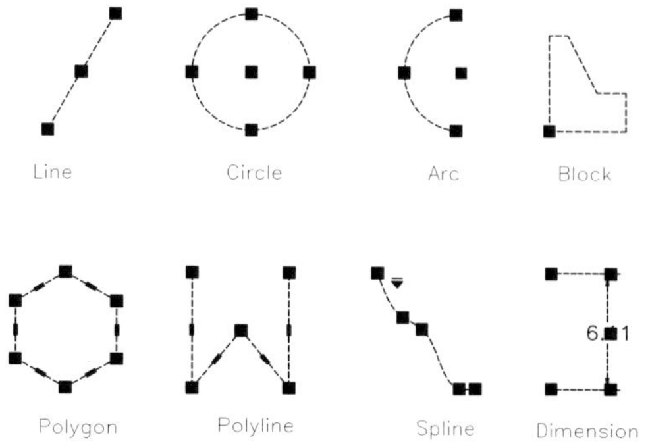

Figure 10-1 *Grips displayed on various objects*

TYPES OF GRIPS

Grips are classified into three types: unselected grips, hover grips, and selected grips. The selected grips are also called hot grips. When you select an object, the grips are displayed at the definition points of the object and the object is displayed as Neon blue line. These grips are called unselected grips and are displayed in blue. When the cursor is moved over an unselected grip, and paused for a second, the grip will be displayed in orange. These grips are called hover grips. Dimensions corresponding to a hover grip are displayed when you place the cursor on the grip.

ADJUSTING GRIP SETTINGS

Application Menu: Options **Command:** OPTIONS/OP

The grip settings can be adjusted by using the options in the **Selection** tab of the **Options** dialog box. This dialog box can also be invoked by choosing **Options** from the shortcut menu displayed upon right-clicking in the drawing area or by choosing the **Options** button from the **Application Menu**.

EDITING OBJECTS BY USING GRIPS

As mentioned earlier, you can perform different kinds of editing operations using the selected grip. The editing operations are discussed next.

Stretching the Objects by Using Grips (Stretch Mode)

If you select an object, the unselected grips are displayed at the definition points of the object. If you select a grip for editing, you are automatically in the **Stretch** mode. The **Stretch** mode has a function similar to that of the **Stretch** tool. When you select a grip, it acts as a base point and is called a base grip. To select several grips, press and hold the SHIFT key and then select the grips. Now, release the SHIFT key and select one of the hot grips and stretch it; all selected grips will be stretched.

Moving the Objects by Using Grips (Move Mode)

The **Move** mode allows you move the selected objects to a new location. When you move objects, their size and angles do not change. You can also use this mode to make the copies of the selected objects or to redefine the base point.

Rotating the Objects by Using Grips (Rotate Mode)

The **Rotate** mode allows you to rotate objects around the base point without changing their size. The options of the **Rotate** mode can be used to redefine the base point, specify a reference angle, or make multiple copies that are rotated about the specified base point. You can access the **Rotate** mode by selecting the grip and then choosing **Rotate** from the shortcut menu, or by giving a null response twice by pressing the SPACEBAR or the ENTER key.

Scaling the Objects by Using Grips (Scale Mode)

The **Scale** mode allows you to scale objects with respect to the base point without changing their orientation. The options of **Scale** mode can be used to redefine the base point, specify a reference length, or make multiple copies that are scaled with respect to the specified base point. You can access the **Scale** mode by selecting the grip and then choosing **Scale** from the shortcut menu, or giving a null response three times by pressing the SPACEBAR or the ENTER key.

Mirroring the Objects by Using Grips (Mirror Mode)

The **Mirror** mode allows you to mirror the objects across the mirror axis without changing their size. The mirror axis is defined by specifying two points. The first point is the base point, and the second point is the point that you select when AutoCAD prompts for the second point.

Editing a Polyline by Using Grips

In AutoCAD, when you select a polyline, three grips are displayed in each segment of the polyline. You can edit the polyline by using these grips. To do so, move the cursor on one of the grips and pause for a while; a tooltip will be displayed, as shown in Figure 10-2. To stretch the polyline, choose the **Stretch Vertex** option and specify a new point; the polyline will be stretched. After choosing the **Stretch Vertex** option, you can also invoke the **Base Point** or **Copy** option from the shortcut menu, as discussed earlier.

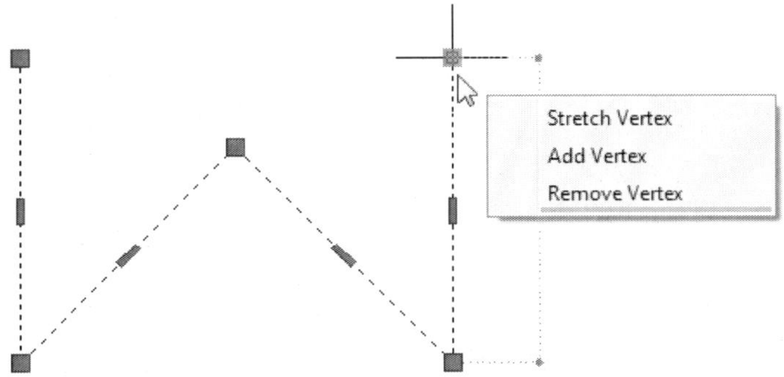

Figure 10-2 *Tooltip displayed near the grip*

EDITING GRIPPED OBJECTS

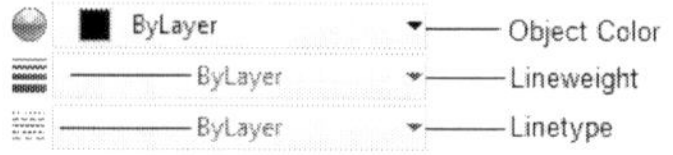

*Figure 10-3 The **Properties** panel*

You can also edit the properties of the gripped objects by using the **Properties** panel in the **Home** tab, refer to Figure 10-3. When you select objects without invoking a tool, the grips (rectangular boxes) will be displayed on the selected objects. For example, to change the color of the gripped objects, select the **Object Color** drop-down list in the **Properties** panel and then select a color from it. The color of the gripped objects will change to the selected color.

CHANGING THE PROPERTIES USING THE PROPERTIES PALETTE

Ribbon: View > Palettes > Properties **Toolbar:** Quick Access > Properties
Command: PROPERTIES/PR/MO/CH

As mentioned earlier, each object has properties associated with it such as the color, layer, linetype, line weight, and so on. You can modify these properties by using the **PROPERTIES** palette. To view this palette, choose the **Properties** tool from the **Palettes** panel in the **View** tab; the **PROPERTIES** palette will be displayed. The **Properties** palette can also be displayed when you double-click on the object to be edited. The contents of the **PROPERTIES** palette change depending upon the objects selected. For example, if you select a text entity, the related properties such as its height, justification, style, rotation angle, and so on, will be displayed.

MATCHING THE PROPERTIES OF SKETCHED OBJECTS

Ribbon: Home > Properties > Match Properties **Command:** MATCHPROP/MA
Quick Access Toolbar: Match Properties (Customize to Add)

The **Match Properties** tool is used to apply properties like color, layer, linetype, and linetype scale of a source object to the selected objects. On invoking this tool, you will be prompted to select the source object and then the destination objects. The properties of the destination objects will be replaced with the properties of the source object. This is a transparent tool and can be used when another tool is active.

QUICK SELECTION OF SKETCHED OBJECTS

Ribbon: Home > Utilities > Quick Select **Command:** QSELECT

The **Quick Select** tool is used to create a new selection set that will either include or exclude all objects whose object type and property criteria match as specified for the selection set. The **Quick Select** tool can be used to select the entities in the entire drawing or in the existing selection set. If a drawing is partially opened, the **Quick Select** tool does not consider the objects that are not loaded. The **Quick Select** tool can be invoked from the **Utilities** panel. You can also choose the **Quick Select** option from the shortcut menu displayed by right clicking in the drawing area.

CYCLING THROUGH SELECTION

In AutoCAD, you can cycle through the objects to be selected if they are overlapping or close to other entities. This feature helps in selecting the entities easily and quickly. To enable this feature, choose the **Selection Cycling** button in the Status Bar. Now, if you move the cursor near an entity that has other entities nearby it then the selection cycling symbol will be displayed.

MANAGING CONTENTS USING THE DesignCenter

Ribbon: View > Palettes > DesignCenter **Command:** ADCENTER/ADC/DC
Menu Bar: Tools > Palettes > DesignCenter

The **DESIGNCENTER** palette is used to locate and organize drawing data, and to insert blocks, layers, external references, and other customized drawing content. You can use the **DesignCenter** to conveniently drag and drop any information that has been previously created into the current drawing. This powerful tool reduces the repetitive tasks of creating information that already exists.

Tutorial 1 DesignCenter

Use the **DesignCenter** to locate and view the contents of the drawing *Kitchens.dwg*. Also, use the **DesignCenter** to insert a block from this drawing and import a layer and a textstyle from the *Blocks and Tables - Imperial.dwg* file located in the **Sample** folder. Use these to make a drawing of a Kitchen plan (*MyKitchen.dwg*) and then add text to it, as shown in Figure 10-4.

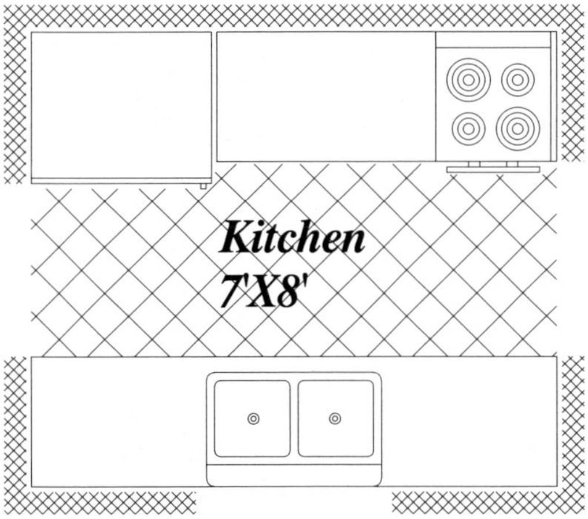

Figure 10-4 *Drawing for Tutorial 1*

1. Open a new drawing using the **Start from Scratch** option. Make sure the **Imperial (feet and inches)** radio button is selected in the **Create New Drawing** dialog box.

2. Change the units to **Architectural** using the **Drawing Units** dialog box. Increase the limits to **10',10'**. Use the **Zoom All** tool to increase the drawing display area.

3. Choose the **DesignCenter** tool from the **Palettes** panel in the **View** tab; the **DesignCenter** window is displayed at its default location.

4. In the **DesignCenter** toolbar, choose the **Tree View Toggle** button to display the Tree View and the Palette (if is not already displayed). Also, choose the **Preview** button, if it is not displayed already. You can resize the window, if needed, to view both the Tree View and the Palette, conveniently.

5. Choose the **Search** button in the **DesignCenter** to display the **Search** dialog box. Here, select **Drawings** from the **Look for** drop-down list and **C:** (or the drive in which AutoCAD 2019 is installed) from the **In** drop-down list. Select the **Search subfolders** check box. In the **Drawings** tab, type **Kitchens** in the **Search for the word(s)** edit box and select **File Name** from the **In the field(s)** drop-down list. Now, choose the **Search Now** button to commence the search. After the drawing has been located, its details and path are displayed in a list box at the bottom of the dialog box.

6. Now, right-click on *Kitchens.dwg* in the list box of the **Search** dialog box and choose **Load into Content Area** from the shortcut menu. You will notice that the drawing and its contents are displayed in the Tree view.

7. Close the **Search** dialog box, if it is still open.

8. Double-click on *Kitchens.dwg* in the Tree View to expand the tree view and display its contents, in case they are not displayed. You can also expand the contents by clicking on the + sign located on the left of the file name in the Tree view.

9. Select **Blocks** in the Tree View to display the list of blocks in the drawing in the **Palette**. Using the left mouse button, drag and drop the block **Kitchen Layout-7x8 ft** in the current drawing.

10. Now, double-click on the *AutoCAD Textstyles and Linetypes.dwg* file located in the **DesignCenter** folder in the same directory to display its contents in the Palette.

11. Double-click on **Textstyles** to display the list of text styles in the Palette. Select **Dutch Bold Italic** in the **Palette** and drag and drop it in the current drawing. You can use this textstyle for adding text to the current drawing.

12. Invoke the **Multiline Text** tool and use the imported textstyle to add the text to the current drawing and complete it, refer to Figure 10-4.

13. Save the current drawing with the name *MyKitchen.dwg*.

Note

1. To create text, refer to Figure 10-4, you need to change the text height.

2. You can import Blocks, Dimstyles, Layers, and so on to the current drawing from any existing drawing.

MAKING INQUIRIES ABOUT OBJECTS AND DRAWINGS

In AutoCAD, you need to use the inquiry tools or commands to obtain information about the selected objects.

Measuring Area of Objects

Ribbon: Home > Utilities > Measure drop-down > Area **Menu Bar:** Tools > Inquiry > Area
Toolbar: Inquiry > Distance drop-down > Area **Command:** AREA/AA

In AutoCAD, the **Area** tool is used to automatically calculate the area of an object in square units. You can use the default option of the **Area** tool to calculate the area and perimeter or circumference of the space enclosed by the sequence of specified points.

Measuring the Distance Between Two Points

Ribbon: Home > Utilities > Measure drop-down > Distance
Toolbar: Inquiry > Distance drop-down
Menu Bar: Tools > Inquiry > Distance **Command:** DIST/DI

The **Distance** tool is used to measure the distance between two selected points. The distance computed by the **Distance** tool is saved in the **DISTANCE** variable.

Listing Information about Objects

Ribbon: Home > Properties > List **Toolbar:** Inquiry > List
Menu Bar: Tools > Inquiry > List **Command:** LIST/LI

The **List** tool displays all the information pertaining to the selected objects. The information is displayed in the **AutoCAD Text Window**.

EXERCISES

Exercise 1

1. Use the **Line** tool to draw the shape shown in Figure 10-5(a).
2. Use grips (**Stretch** mode) to get the shape shown in Figure 10-5(b).
3. Use the **Rotate** and **Stretch** modes to get the copies shown in Figure 10-5(c).

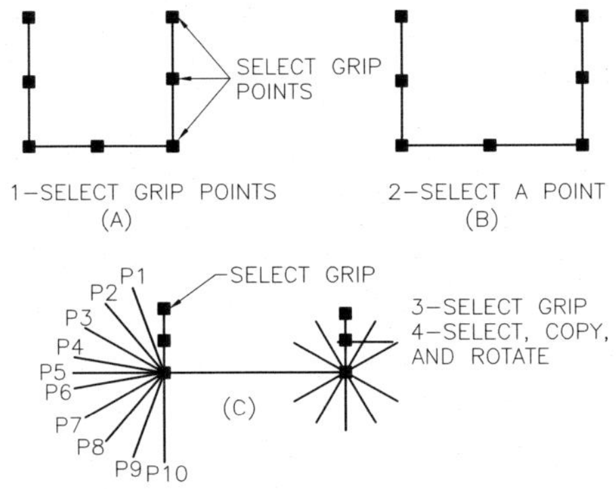

Figure 10-5 *Drawing for Exercise 1*

Exercise 2

Use the drawing and editing tools to create the sketch shown in Figure 10-6.

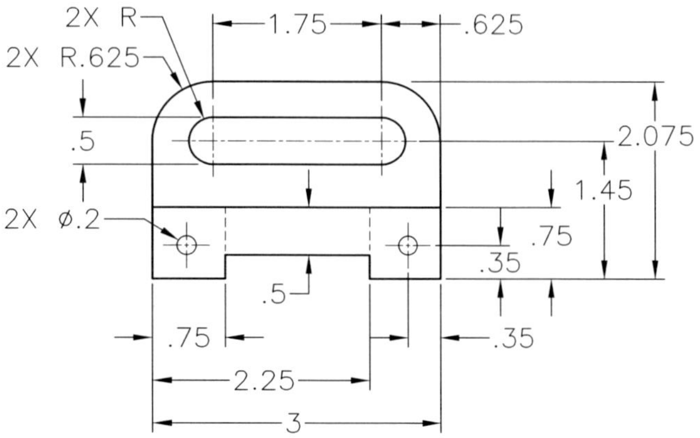

Figure 10-6 Drawing for Exercise 2

Exercise 3

Use the drawing and editing tools to create the sketch shown in Figure 10-7.

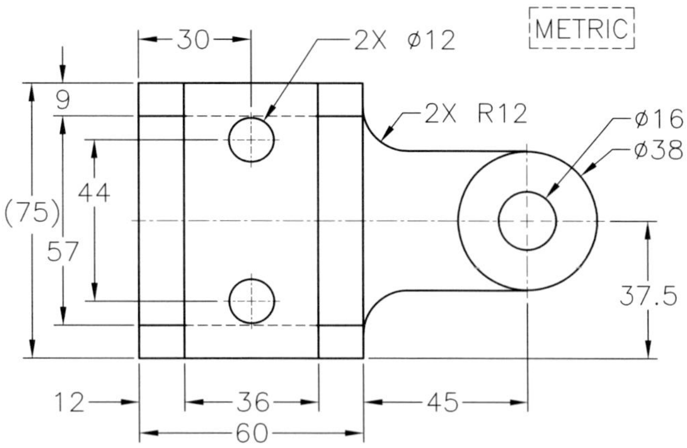

Figure 10-7 Drawing for Exercise 3

Chapter 11

Adding Constraints to Sketches

Learning Objectives

After completing this chapter, you will be able to:

- *Add geometric constraints to sketches*
- *Control the display of constraints*
- *Apply constraints automatically*
- *Apply dimensional constraints to sketches*

INTRODUCTION

In AutoCAD, a wide set of tools are available that help you create parametric drawings. These tools are available in the **Parametric** tab of the **Ribbon** as well as in the **Parametric** toolbar. The parametric nature of a drawing implies that the shape and size of a geometry can be defined using standard properties or parameters. The main function of parametric property is to drive a new size or shape to the selected geometry without considering its original dimensions. Also, you can modify the shape and size of any entity of a sketch at any stage of the design process. This property makes the designing process very easy.

ADDING GEOMETRIC CONSTRAINTS

AutoCAD provides you with twelve types of geometric constraints. Usually, these constraints are applied between two entities in a sketch and control the shape of a drawing.

Applying the Horizontal Constraint

Ribbon: Parametric > Geometric > Horizontal **Command:** GEOMCONSTRAINT > H
Toolbar: Geometric Constraint > Horizontal or
Parametric > Constraint drop-down > Horizontal
Menu Bar: Parametric > Geometric Constraints > Horizontal

The Horizontal constraint forces a selected line, polyline, or two points to become horizontal. To apply this constraint, choose the **Horizontal** tool from the **Geometric** panel of the **Parametric** tab of the **Ribbon**.

Applying the Vertical Constraint

Ribbon: Parametric > Geometric > Vertical **Command:** GEOMCONSTRAINT > V
Toolbar: Geometric Constraint > Vertical or Parametric > Constraint drop-down > Vertical
Menu Bar: Parametric > Geometric Constraints > Vertical

The Vertical constraint forces a selected line, polyline, or two points to become vertical. To apply this constraint, choose the **Vertical** tool from the **Geometric** panel of the **Parametric** tab.

Applying the Coincident Constraint

Ribbon: Parametric > Geometric > Coincident **Command:** GEOMCONSTRAINT > C
Toolbar: Geometric Constraint > Coincident or
Parametric > Constraint drop-down > Coincident
Menu Bar: Parametric > Geometric Constraints > Coincident

The Coincident constraint forces a selected point or keypoint of an entity to be coincident with the point or keypoint of another entity. The entity selected can be a line, polyline, circle, an arc or ellipse. Similarly, the selected point can be a point, or keypoints such as an endpoint or midpoint of a line, an arc or a polyline, or the centerpoint of an arc, ellipse, or a circle. To apply this constraint, choose the **Coincident** tool from the **Geometric** panel of the **Parametric** tab; the cursor will be replaced by a selection box along with the coincident symbol. Also, you will be prompted to select first point. Move the cursor over an entity; the keypoint that is close to the cursor will be highlighted in red. Next, click to select it; you will be prompted to select the second point to be coincident with the first object. Select the keypoint of the second entity; the keypoint of the second entity will move and coincide with that of the first entity.

Applying the Fix Constraint

Ribbon: Parametric > Geometric > Fix **Command:** GEOMCONSTRAINT > F
Toolbar: Geometric Constraint > Fix or Parametric > Constraint drop-down > Fix
Menu Bar: Parametric > Geometric Constraints > Fix

The Fix constraint forces the selected entity to become fixed at a position. If you apply this constraint to a line or to an arc, its location will be fixed with respect to the selected keypoint. But you can change the size of the entity by dragging its endpoints. To apply the Fix constraint, choose the **Fix** tool from the **Geometric** panel of the **Parametric** tab; you will be prompted to select a point or an object. Next, move the cursor over the entity; the point closer to the cursor will be highlighted. Click to fix the highlighted keypoint. Now, you cannot move the fixed keypoint. However, you can drag and relocate other keypoints.

Applying the Perpendicular Constraint

Ribbon: Parametric > Geometric > Perpendicular **Command:** GEOMCONSTRAINT > P
Toolbar: Geometric Constraint > Perpendicular or
Parametric > Constraint drop-down > Perpendicular
Menu Bar: Parametric > Geometric Constraints > Perpendicular

The Perpendicular constraint forces the selected lines to become perpendicular to each other. To apply this constraint, choose the **Perpendicular** tool from the **Geometric** panel; you will be prompted to select the first object. Select a line or a polyline; you will be prompted to select the second object. Select another line or a polyline. The object selected second will become perpendicular to the object selected first.

Applying the Parallel Constraint

Ribbon: Parametric > Geometric > Parallel **Command:** GEOMCONSTRAINT > PA
Toolbar: Geometric Constraint > Parallel or Parametric > Constraint drop-down > Parallel
Menu Bar: Parametric > Geometric Constraints > Parallel

The Parallel constraint forces the selected lines to become parallel to each other. To apply this constraint, choose the **Parallel** tool from the **Geometric** panel; you will be prompted to select the first object. Select a line or a polyline; you will be prompted to select the second object. Select another line or a polyline; the object selected second will become parallel to the object selected first.

Applying the Collinear Constraint

Ribbon: Parametric > Geometric > Collinear **Command:** GEOMCONSTRAINT > COL
Toolbar: Geometric Constraint > Collinear or Parametric > Constraint drop-down > Collinear
Menu Bar: Parametric > Geometric Constraints > Collinear

The Collinear constraint forces the selected lines to lie on the same infinite line. To apply this constraint, choose the **Collinear** tool from the **Geometric** panel; you will be prompted to select the first object. Select a line; you will be prompted to select the second object. Select another line; the line selected later will become collinear with the first line.

Applying the Concentric Constraint

Ribbon: Parametric > Geometric > Concentric **Command:** GEOMCONSTRAINT > CON
Toolbar: Geometric Constraint > Concentric or
Parametric > Constraint drop-down > Concentric
Menu Bar: Parametric > Geometric Constraints > Concentric

The Concentric constraint forces the selected arc or circle to share the center point of another arc, circle, or ellipse. To apply this constraint, choose the **Concentric** tool from the **Geometric** panel; you will be prompted to select the first object. Select an arc, ellipse, or a circle; you will be prompted to select the second object. Select another arc, ellipse, or circle; the object selected second will become concentric to the object selected first.

Applying the Tangent Constraint

Ribbon: Parametric > Geometric > Tangent **Command:** GEOMCONSTRAINT > T
Toolbar: Geometric Constraint > Tangent or Parametric > Constraint drop-down > Tangent
Menu Bar: Parametric > Geometric Constraints > Tangent

The Tangent constraint forces the selected arc, circle, spline, or ellipse to become tangent to another arc, circle, spline, ellipse, or line. To apply this constraint, choose the **Tangent** tool from the **Geometric** panel; you will be prompted to select the first object. Select an arc, circle, spline, line, or an ellipse; you will be prompted to select the second object. Select another arc, circle, spline, ellipse, or line; the object selected second will become tangent to the object selected first.

Applying the Symmetric Constraint

Ribbon: Parametric > Geometric > Symmetric **Command:** GEOMCONSTRAINT > S
Toolbar: Geometric Constraint > Symmetric or
Parametric > Constraint drop-down > Symmetric
Menu Bar: Parametric > Geometric Constraints > Symmetric

The Symmetric constraint forces two selected lines, arcs, circles, or ellipses to remain equidistant from the centerline. To apply this constraint, choose the **Symmetric** tool from the **Geometric** panel; you will be prompted to select the first object or two points. Select an arc, an ellipse, a line, or a circle; you will be prompted to select the second object. Select another arc, ellipse, circle, or line; you will be prompted to select the symmetry line. Select a line; the selected objects will become symmetric with respect to the selected line.

Applying the Equal Constraint

Ribbon: Parametric > Geometric > Equal **Command:** GEOMCONSTRAINT > E
Toolbar: Geometric Constraint > Equal or Parametric > Constraint drop-down > Equal
Menu Bar: Parametric > Geometric Constraints > Equal

The Equal constraint forces the selected lines or polylines to have equal length, or the selected circles, arcs, or an arc and a circle to have equal radii. To apply this constraint, choose the **Equal** tool from the **Geometric** panel; you will be prompted to select an object. Select the first object; you will be prompted to select the second object. Select another object; the size of the second object will become equal to that of the object selected first.

Applying the Smooth Constraint

Ribbon: Parametric > Geometric > Smooth **Command:** GEOMCONSTRAINT > SM
Toolbar: Geometric Constraint > Smooth or Parametric > Constraint drop-down > Smooth
Menu Bar: Parametric > Geometric Constraints > Smooth

The Smooth constraint is used to apply curvature continuity between a spline and an entity connected together. The entities that can be connected to a spline include a line, spline, polyline, or an arc. To apply this constraint, choose the **Smooth** tool from the **Geometric** panel; you will be prompted to select the first object. Select a spline as the first object; you will be prompted to select the second object. Select another object; the smooth constraint will be applied between the selected objects.

CONTROLLING THE DISPLAY OF CONSTRAINTS

To control the display of the constraints, click on the inclined arrow in the **Geometric** panel of the **Parametric** tab; the **Constraint Settings** dialog box will be displayed. In this dialog box, the **Geometric** tab will be chosen and the check boxes of all constraints will be selected. Also, the **Show constraint bars after applying constraints to selected objects** check box will be selected by default. Therefore, on applying a constraint, a constraint bar will be displayed above the sketch and the symbol of that particular constraint will be displayed in the constraint bar by default. If you do not need any particular constraint symbol to be displayed then in the **Constraint bar display settings** area, clear the check box of the constraints that you do not want to be displayed. If you do not need the constraint bar to be displayed on applying the constraints, then clear the **Show constraint bars after applying constraints to selected objects** check box. However, if the **Show constraint bars when objects are selected** check box is selected, then the constraint bars will be displayed when you select an object.

APPLYING CONSTRAINTS AUTOMATICALLY

In the previous section, twelve types of constraints are discussed. Nine of these constraints can be applied automatically, provided you have a closed sketch. To apply constraints automatically, choose the **AutoConstrain** tool in the **Geometric** panel of the **Parametric** tab of the **Ribbon**; you will be prompted to select an object. Select the entities individually or by using the selection box, and then press ENTER; all possible constraints will be applied automatically.

APPLYING DIMENSIONAL CONSTRAINTS

In AutoCAD, you can apply dimensional constraint to an entity and change the size of the entity with ease. You can apply the horizontal, vertical, aligned, angular, radius, and diameter dimensional constraint to a sketch. The procedure to apply a dimensional constraint to an entity is similar to that of applying dimension to an entity.

Tutorial 1

Draw the sketch shown in Figure 11-1 and apply appropriate constraints to it. Do not dimension the sketch.

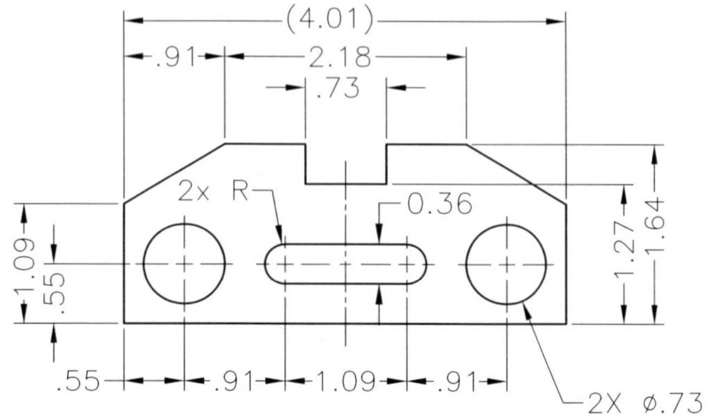

Figure 11-1 *Sketch for Tutorial 1*

In this tutorial, you will select the **Infer geometric constraints** check box and draw the sketch. Therefore, some of the constraints will be applied while sketching. Next, you will apply the constraints that are not applied automatically while sketching. First, you will create the outer loop of the sketch using the **Line** tool.

1. Start a new file with the *acad.dwt* template in the **Drafting & Annotation** workspace.

2. Ensure that the **Infer Constraints** button is chosen in the Status Bar. By using the **Line** tool, draw the outer profile similar to that shown in Figure 11-1. Make sure that the endpoints are snapped so that they form a closed sketch, as shown in Figure 11-2. Note that in your drawing, the constraints applied by default may be different from that shown in Figure 11-2, as it depends upon the start point and the procedure used to draw the lines.

   Next, you need to apply the constraints that are not applied by default. First, you will apply the Equal constraint.

3. Choose the **Equal** tool from the **Geometric** panel in the **Parametric** tab; the cursor changes to a selection box with the Equal constraint's symbol attached to it. Also, you are prompted to select the first object.

4. Select the vertical line on the left as the first object; you are prompted to select the second object.

5. Select the vertical line on the right as the second object; the equal constraint is applied and the constraint bar is displayed.

6. Similarly, apply the equal constraint between the two inclined lines, two small vertical lines, and two small horizontal lines. The sketch after applying all the constraints is shown in Figure 11-3.

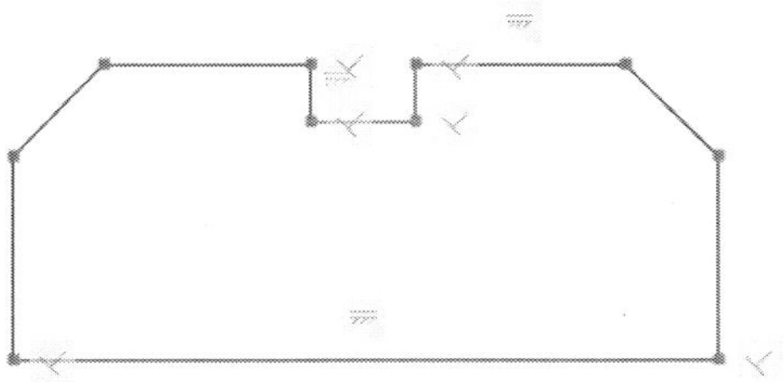

Figure 11-2 *Outer loop of the sketch*

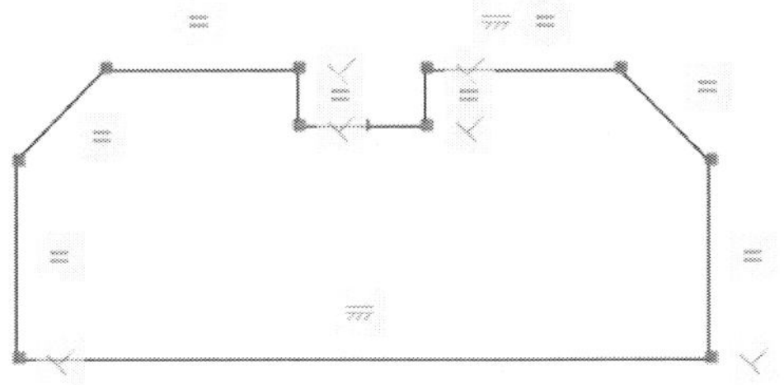

Figure 11-3 *Sketch after applying all constraints*

Next, you need to apply the symmetric constraint between entities on either side of the centerline.

7. Choose the **Infer Constraints** button from the Status Bar to turn off the infer constraints.
8. Draw a vertical centerline that passes through the midpoint of the horizontal line at the bottom, as shown in Figure 11-4.
9. Choose the **Symmetric** tool from the **Geometric** panel; the cursor changes to a selection box with symmetric symbol attached to it. Also, you are prompted to select the first object.
10. Select the vertical line on the left as the first object; you are prompted to select the second object.
11. Select the vertical line on the right as the second object; you are prompted to select the symmetry line.
12. Select the centerline as the symmetry line; the vertical lines become symmetric to the centerline and the symmetric symbol is displayed, as shown in Figure 11-5. Note that in this figure, other constraints are hidden for clarity.

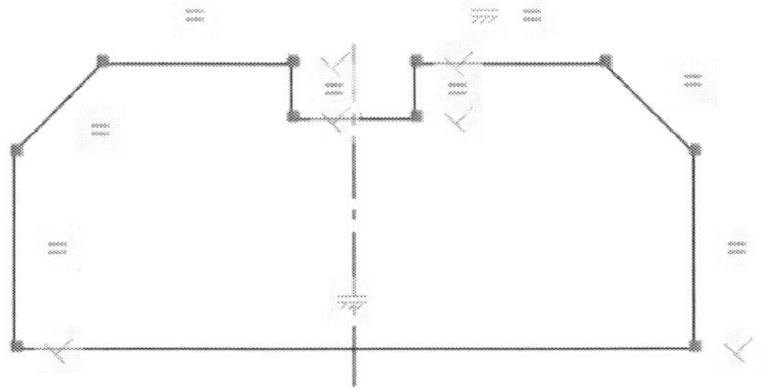

Figure 11-4 *Sketch after drawing the centerline*

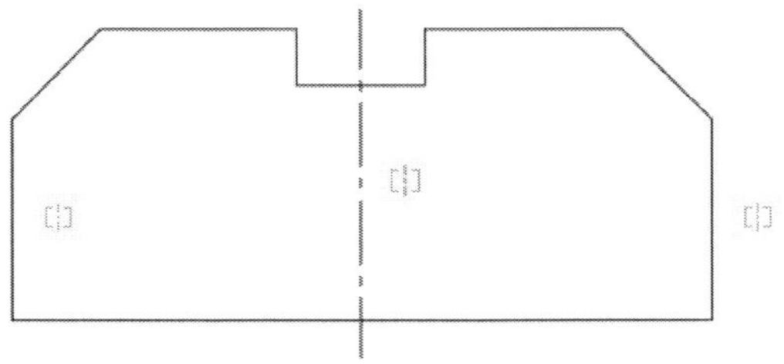

Figure 11-5 *The Symmetric constraint symbol*

13. Similarly, apply the symmetric constraints between the two inclined lines and then between the two small horizontal lines. The three symmetric constraints are applied as shown in Figure 11-6.
14. Choose the **Hide All** button from the **Parametric** tab of the **Geometric** panel to hide all constraint bars. Figure 11-7 shows the sketch after hiding all constraint bars.

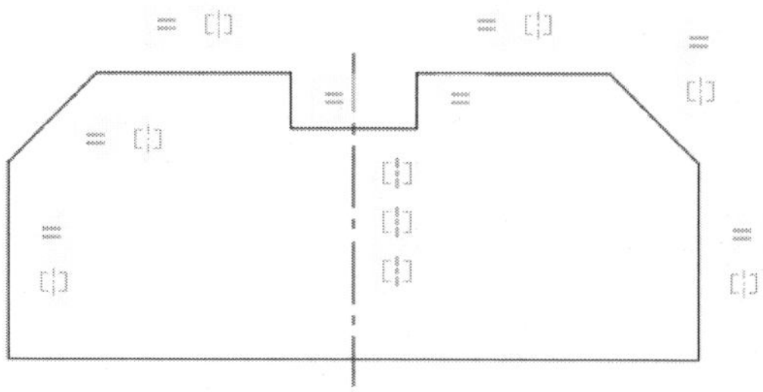

Figure 11-6 *Sketch with symmetric constraints*

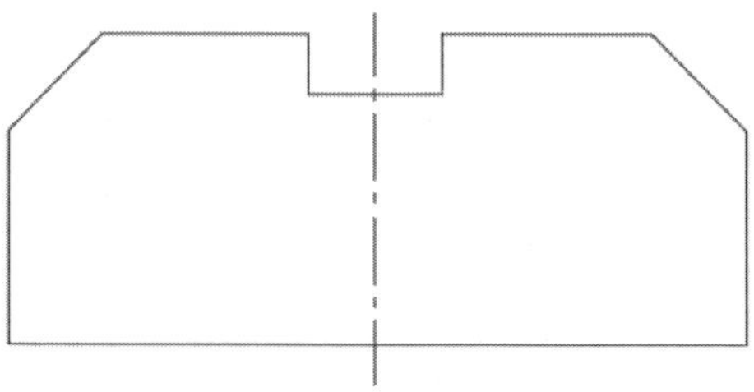

Figure 11-7 *Sketch after hiding all constraints*

Next, you need to create the inner loop using the **Arc**, **Circle**, and **Line** tools at an arbitrary location. But the profile should be similar to that given in Figure 11-1. Next, you need to apply the constraints.

15. Draw two circles of any radius inside the outer loop.

16. Choose the **Infer Constraints** button from the Status Bar to turn on the infer constraints.

17. Draw a slot using the **Center**, **Start**, **End** arc and **Line** tools. The sketch after drawing the circles and the slot is shown in Figure 11-8. You can also draw the profile of the slot using the **Polyline** tool. However, the sketch in this tutorial has been drawn using the **Center**, **Start**, **End** arc and **Line** tools.

18. Apply the equal constraint first between two circles and then between two arcs.

19. Apply the symmetric constraint between the circles as well as between the arcs.

20. Choose the **Tangent** tool from the **Geometric** panel; the cursor changes to a selection box with the tangent symbol attached to it. Also, you are prompted to select the first object.

21. Select an arc; you are prompted to select the second object.

22. Select a line; the tangent constraint will be applied at one end of the line and the tangent symbol will be displayed.

23. Similarly, apply the tangent constraint at all other places in the slot, as shown in Figure 11-9. You need to choose the **Show/Hide** button to display all constraints.

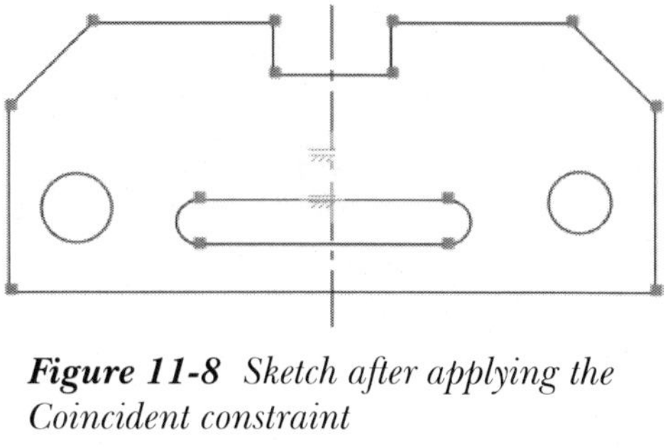

Figure 11-8 *Sketch after applying the Coincident constraint*

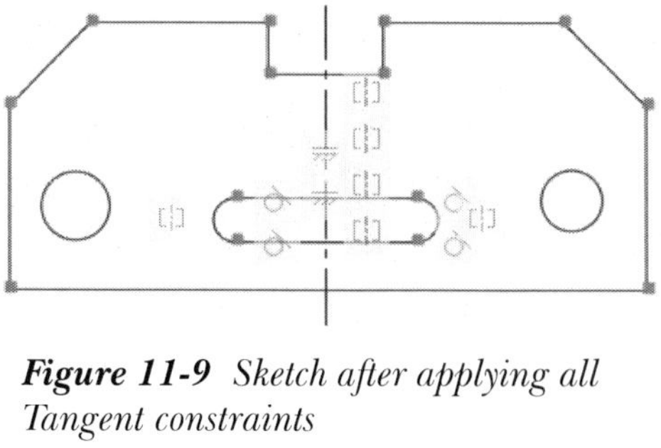

Figure 11-9 *Sketch after applying all Tangent constraints*

Next, you need to make the centerpoint of the circles and the slot to be in a straight line.

24. Choose the **Horizontal** tool from the **Geometric** panel; the prompt sequence that follows is given next.

    Select an object or [2Points] <2Points>: Enter
    Select first point: *Select the circle on the left side* Enter
    Select second point: *cen* Enter
    of: *Select the center of the arc on the left side* Enter

    The sketch after applying the Horizontal constraint is shown in Figure 11-10.

25. Similarly, apply the Horizontal constraint to the circle and arc on the right side.

26. Choose the **Show/Hide** button from the **Parametric** tab and hide the display of all constraint bars and save the drawing. The sketch after applying and hiding all constraints is shown in Figure 11-11.

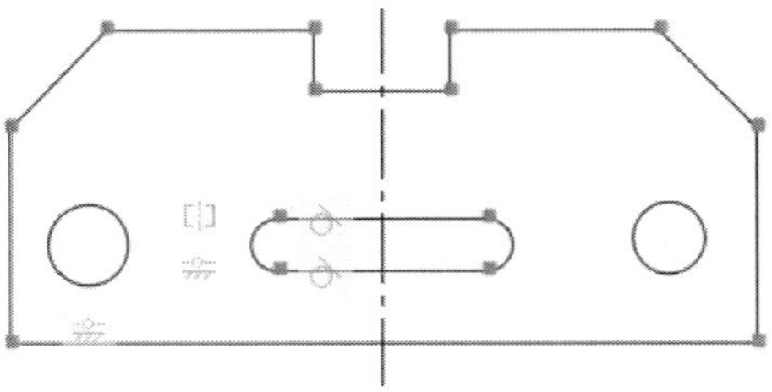

Figure 11-10 *Sketch after applying the Horizontal constraint between a circle and an arc*

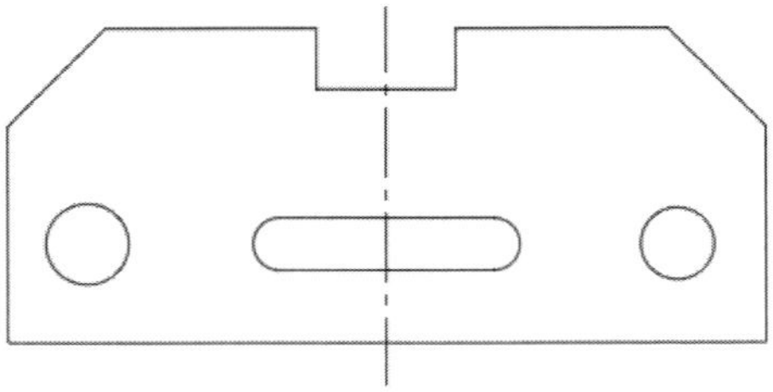

Figure 11-11 *Sketch after applying and hiding all constraints*

Tutorial 2

In this tutorial, you will draw the sketch shown in Figure 11-12, and then apply the geometric and dimensional constraints to it.

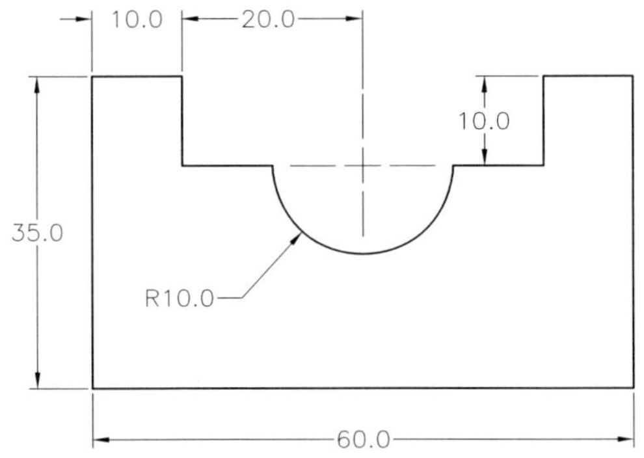

Figure 11-12 *Sketch for Tutorial 2*

First, you will draw the profile and apply the geometrical constraints.

1. Start a new file with the *acad.dwt* template in the **Drafting & Annotation** workspace and draw the outer profile with dimensions close to that given in Figure 11-12. Make sure that the endpoints are snapped so that they form a closed sketch, as shown in Figure 11-13.

2. Choose the **AutoConstrain** button from the **Geometric** panel of the **Parametric** tab, enter **S** at the command prompt, and press ENTER; the **Constraint Settings** dialog box is displayed.

3. Make sure that all constraints are selected. As there is no need to apply the concentric or tangent constraints in this sketch, select the **Concentric** and **Tangent** constraints individually and move them down using the **Move Down** button.

4. Choose the **OK** button; you are prompted to select objects. Select all entities of the sketch by dragging a selection box. Next, press ENTER; all possible constraints are applied to the sketch, as shown in Figure 11-14.

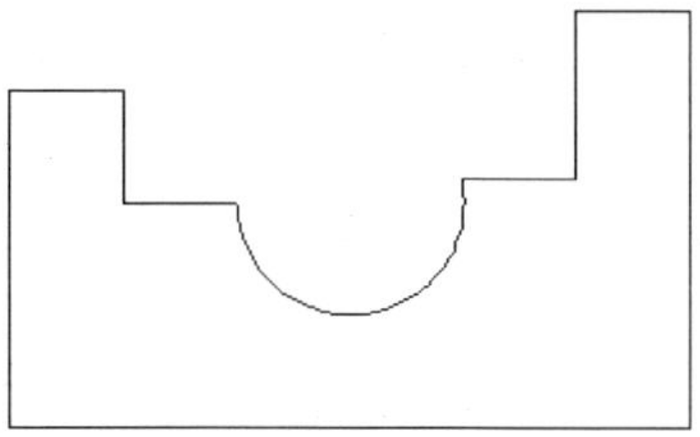

Figure 11-13 *Closed sketch drawn*

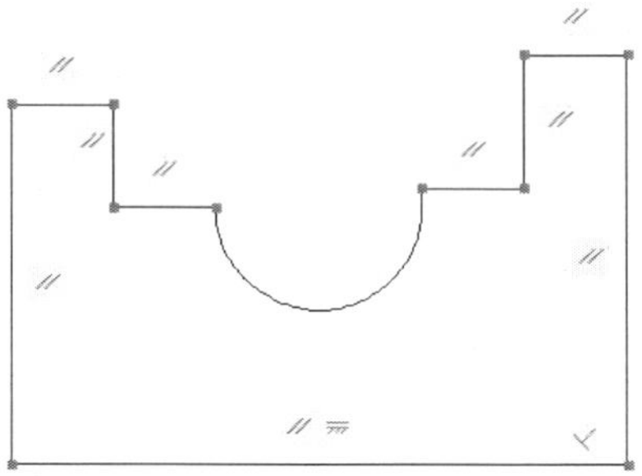

Figure 11-14 *Constraints applied automatically*

Note

*The constraints applied using the **AutoConstrain** tool may vary depending upon the profile. In your case, if the constraints are different from that shown in Figure 11-14, apply the missing constraints. Also, choose the **Show All** button from the **Geometric** panel if the constraints are not visible by default.*

5. Now, you need to apply the equal constraint. To do so, choose the **Equal** tool from the **Geometric** panel of the **Parametric** tab and apply the Equal constraint between the lines that are supposed to have equal dimensions. For dimensions, refer to Figure 11-12. The equal constraints need to be applied to the two pairs of horizontal and two pairs of vertical lines in the sketch. The sketch after applying the Equal constraint is shown in Figure 11-15.

Note

If you do not want to apply the Equal constraint to vertical lines, you need to apply the Collinear constraint to the two pairs of horizontal lines. But after applying the Equal constraint if you apply the Collinear constraint, a message box will be displayed stating that this constraint cannot be applied and conflicts the existing constraints or is over-constraining the geometry.

Next, you need to make the centerpoint of the circle and the horizontal line to be in a line.

6. Choose the **Horizontal** tool from the **Geometric** panel of the **Parametric** tab; the prompt sequence that will follow is given next.

   Select an object or [2Points] <2Points>: Enter
   Select first point: *Move the cursor over the horizontal line and click when the endpoint near the arc is highlighted*
   Select second point: *cen* Enter *and select the arc on the left side*

   The sketch after applying the **Horizontal** constraint is shown in Figure 11-16.

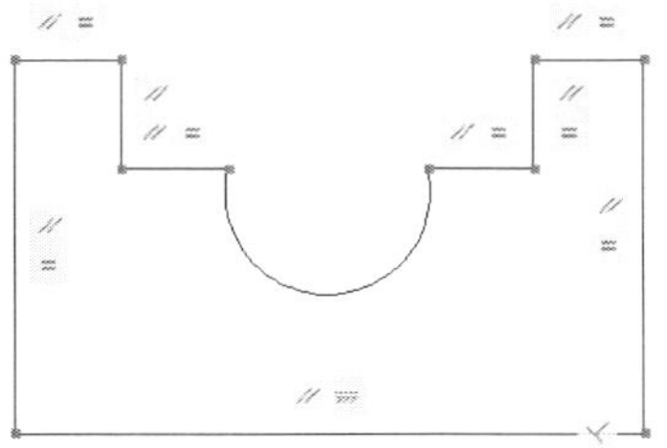

Figure 11-15 *Sketch after applying the Equal constraint*

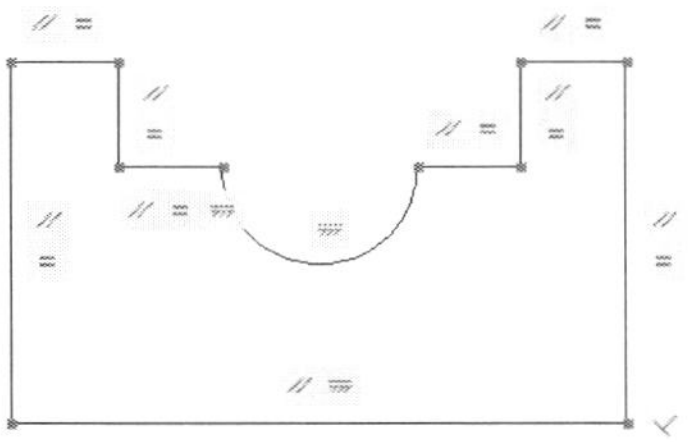

Figure 11-16 *Sketch after applying the Horizontal constraint*

7. Choose the **Hide All** button from the **Geometric** panel in the **Parametric** tab; all constraint bars disappear.

Now, you need to apply dimensional constraints. Before that, expand the **Dimensional** panel and ensure that the **Dynamic Constraint Mode** option is chosen.

8. Choose the **Linear** tool from **Parametric > Dimensional > Dimensional Constraint** drop-down and follow the prompt sequence given next.

   Specify first constraint point or [Object] <Object>: Enter
   Select object: *Select the horizontal line at the bottom.*
   Specify dimension line location: *Move the cursor down and place the horizontal dimension at the required location. Type **60** in the text box and press ENTER.*

9. Press the ENTER key again or choose the **Linear** tool from the **Dimensional** panel of the **Parametric** tab and follow the prompt sequence given next.

   Specify first constraint point or [Object] <Object>: Enter
   Select object: *Select the vertical line on the left*
   Specify dimension line location: *Move the cursor on the left and place the vertical dimension at the required location. Enter **35** in the textbox.*

10. Similarly, apply the dimension to the horizontal and vertical lines, refer to Figure 11-17 and change the value to **10**.

11. Choose the **Radius** tool from the **Dimensional** panel of the **Parametric** tab; you are prompted to select the arc or the circle.

12. Select the arc and enter **10** as radius.

Now, you need to apply the dimension between the center of the arc and the vertical line on the left, refer to Figure 11-12.

13. Invoke the **Linear** tool from the **Parametric > Dimensional > Dimensional Constraints** drop-down and follow the prompt sequence given next.

    Specify first constraint point or [Object] <Object>: *Select the keypoint at the top of the smaller vertical line on the left.*
    Specify second constraint point: *cen* Enter *and select the arc.*
    Specify dimension line location: *Place the dimension at the required location and enter **20**; the **Dimensional Constraints** message box is displayed.*

14. Choose the **Create a reference dimension** option from the message box. The value **20** is displayed in parenthesis stating that it is a reference dimension. The sketch after applying all dimensional constraints is shown in Figure 11-17.

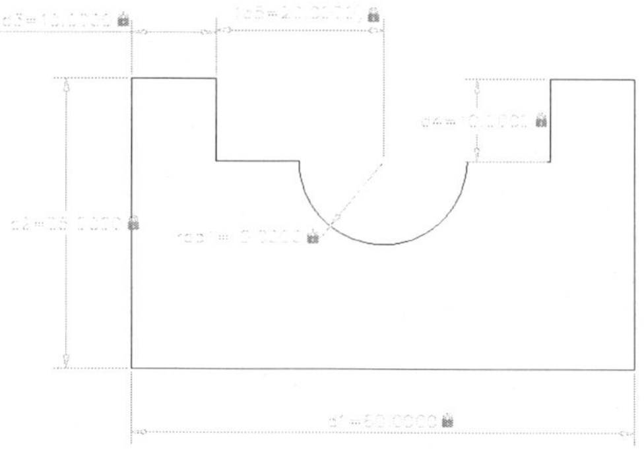

Figure 11-17 *Sketch with all dimensional constraints*

EXERCISES

Exercise 1

Draw the sketch shown in Figure 11-18 and apply appropriate constraints to it. Do not dimension the sketch.

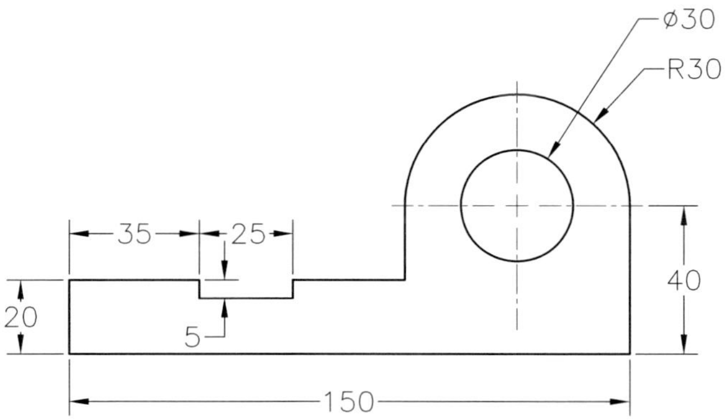

Figure 11-18 *Sketch for Exercise 1*

Exercise 2

Draw the sketches shown in Figure 11-19 and Figure 11-20. Next, add geometric and dimensional constraints to them.

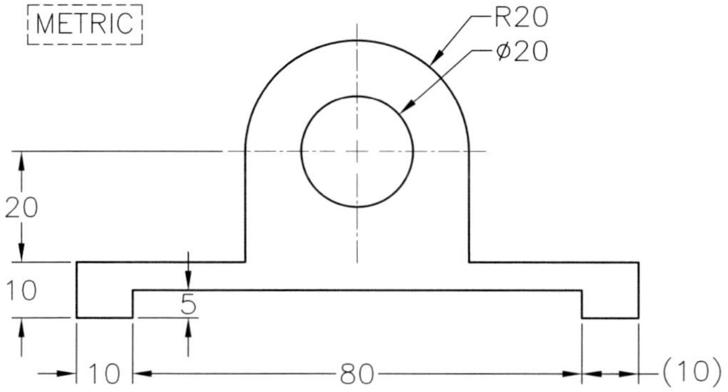

Figure 11-19 *Sketch for Exercise 2*

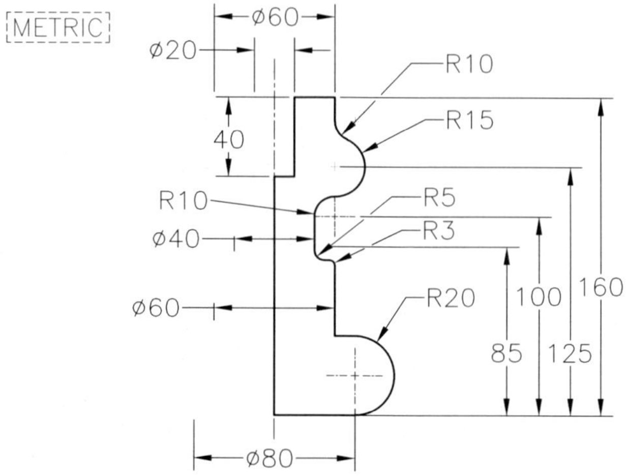

Figure 11-20 *Sketch for Exercise 2*

Chapter 12

Hatching Drawings

Learning Objectives

After completing this chapter, you will be able to:

- *Hatch an area by using the Hatch tool*
- *Specify pattern properties*
- *Preview and apply hatching*
- *Create annotative hatching*

HATCHING

In many drawings, such as sections of solids, the sectioned area needs to be filled with some pattern. Different filling patterns make it possible to distinguish between different parts or components of an object. Also, the material of which an object is made can be indicated by the filling pattern. You can also use these filling patterns in graphics for rendering architectural elevations of buildings, or indicating the different levels in terrain and contour maps. Filling objects with a pattern is known as hatching, refer to Figure 12-1. This hatching process can be accomplished by using the **Hatch** tool in the **Draw** panel of the **Home** tab or the **Tool Palettes**.

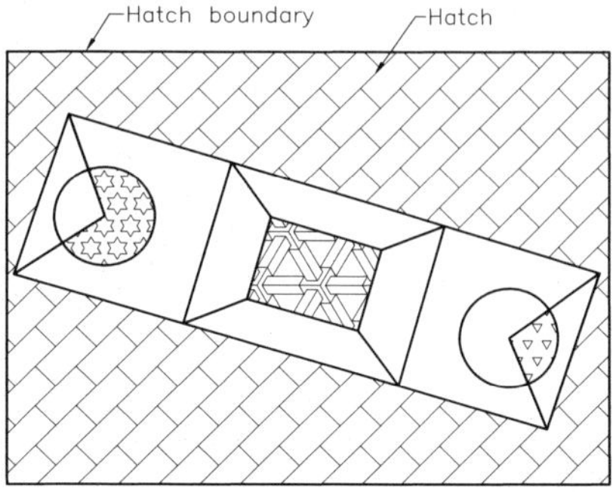

Figure 12-1 *Illustration of hatching*

Before using the **Hatch** tool, you need to understand the terminology related to hatching. Some of the terms are explained next.

Hatch Patterns

AutoCAD supports a variety of hatch patterns. Every hatch pattern consists of one or more hatch lines or a solid fill. The lines are placed at specified angles and spacing. You can change the angle and the spacing between the hatch lines. These lines may be broken into dots and dashes, or may be continuous, as required. The hatch pattern is trimmed or repeated, as required, to fill the specified area exactly.

Hatch Boundary

Hatching can be used on parts of a drawing enclosed by a boundary. This boundary may be lines, circles, arcs, polylines, 3D faces, or other objects, and at least part of each bounding object must be displayed within the active viewport.

HATCHING DRAWINGS USING THE HATCH TOOL

Ribbon: Home >Draw > Hatch drop-down > Hatch	**Toolbar:** Draw > Hatch
Menu Bar: Draw > Hatch	**Command:** HATCH/H

The **Hatch** tool is used to hatch a region enclosed within a boundary (closed area) by selecting a point inside the boundary or by selecting the objects to be hatched. This tool automatically designates a boundary and ignores other objects (whole or partial) that may not be a part of this boundary. When you choose the **Hatch** tool, the **Hatch Creation** tab will be displayed. Also, you will be prompted to pick internal point. Select the type of hatch pattern from the **Pattern** panel of the **Hatch Creation** tab, set the properties of the hatch pattern in the

Properties panel, and move the cursor inside a closed profile; the preview of the hatch pattern will be displayed. Now, pick the internal point; the hatch will be applied.

Tutorial 1 — *Hatch*

In this tutorial, you will hatch a circle using the default hatch settings. Later, in the chapter, you will learn to change the settings to get the desired hatch pattern.

1. Create a circle and then choose the **Hatch** tool from the **Draw** panel; the **Hatch Creation** tab is displayed and you are prompted to **Pick internal point or [Select objects/Undo/seTtings]:** move the cursor inside the circular region; a preview of the hatch pattern is displayed.

2. Select a point inside the circle (P1) (Figure 12-2) and press ENTER; the hatch is applied, as shown in Figure 12-3.

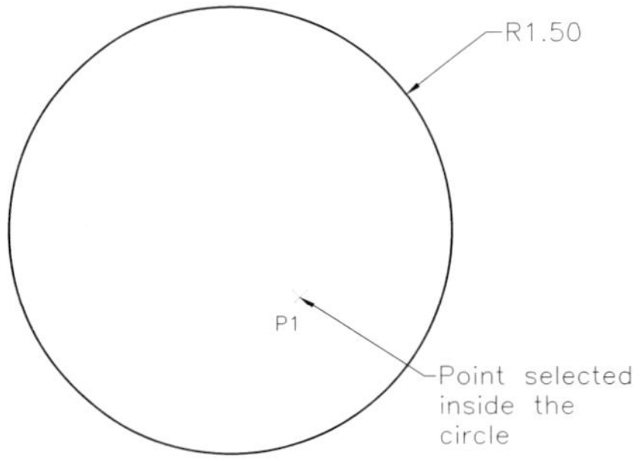

Figure 12-2 *Specifying a point to hatch the circle*

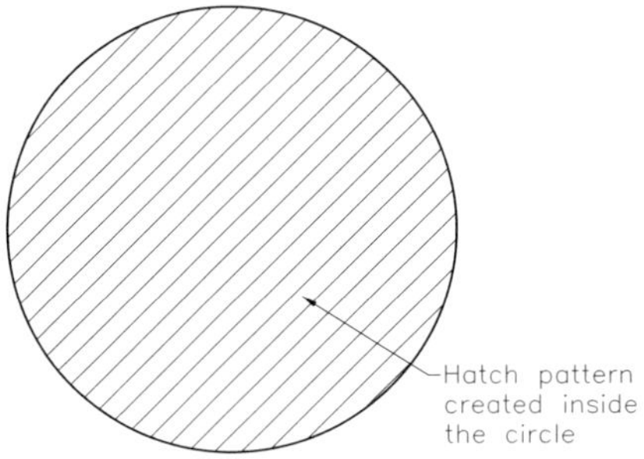

Figure 12-3 *Drawing after hatching*

PANELS IN THE HATCH CREATION TAB

Before or after specifying the pick points, you can change the parameters of a hatch pattern using various options in the panels of the **Hatch Creation** tab. These panels and their options are discussed next.

Boundaries Panel

The options in the **Boundaries** panel are used to define the hatch boundary. This is done by selecting a point inside a closed area to be hatched or by selecting the objects. You can also remove islands, create hatch boundary, and define the boundary set by using the options in this panel, as discussed next.

Pattern Panel

The **Pattern** panel displays all predefined patterns available in AutoCAD. However, the list depends upon the option selected in the **Hatch Type** drop-down list in the **Properties** panel. By default, the **Pattern** option is selected in this drop-down list and the predefined patterns are listed in the **Pattern** panel. A predefined pattern consists of **ANSI**, **ISO**, and other pattern types. The hatch pattern selected in this panel will be applied to the object. The selected pattern will be stored in the **HPNAME** system variable. **ANSI31** is the default pattern in the **HPNAME** system variable. If you need to hatch a solid by using a solid color, then choose the **Solid** option and specify the color in the **Hatch Color** option in the **Properties** panel. Similarly, if you need to create a user-defined pattern, choose the **USER** option from the **Pattern** panel and set the properties.

Properties Panel

The options in this panel are used to set properties of the pattern selected in the **Pattern** panel.

Origin Panel

Hatch pattern alignment is an important feature of hatching, as on many occasions, you need to hatch adjacent areas with similar or sometimes identical hatch patterns while keeping the adjacent hatch patterns properly aligned. Proper alignment of hatch patterns is taken care of automatically by generating all lines of every hatch pattern from the same reference point. The reference point is normally at the origin point (0,0). The options in the **Origin** panel allow you to specify the origin of hatch so that they get properly aligned,

Options Panel

The options in this panel allow you to specify the draw order and some commonly used properties of the hatch pattern.

Setting the Parameters for Gradient Pattern

During the hatching, you can select the **Gradient** option in the **Hatch Type** drop-down list to fill the boundary in a set pattern of colors. On selecting the **Gradient** option from the **Hatch Type** drop-down list, nine fixed patterns will be listed in the **Pattern** panel and their corresponding option will be displayed in the **Hatch Creation** tab. You can select the required gradient or specify the gradient pattern by entering its value in the **GFNAME** system variable. The default value of this variable is 1. As a result, the first gradient pattern is selected.

You can set two colors for the gradient from the **Gradient Color 1** and **Gradient Color 2** drop-down lists available in the **Properties** panel of the **Hatch Creation** tab.

CREATING ANNOTATIVE HATCH

You can create annotative hatch having similar annotative properties like text and dimensions. The annotative hatch are defined relative to the viewport scaling, you only have to specify the hatch scale according to its display on the sheet. The display size of the hatch in the model space will be controlled by the current annotation scale multiplied by the paper space height.

HATCHING THE DRAWING USING THE TOOL PALETTES

Ribbon: View > Palettes > Tool Palettes **Command:** TOOLPALETTES
Toolbar: Standard Annotation > Tool Palettes Window
Menu Bar: Tools > Palettes > Tool Palettes

You can use the **Tool Palettes** to insert predefined hatch patterns and blocks in the drawings. A number of tabs such as **Command Tool Samples, Hatches and Fills**, **Civil,**

Structural, Electrical, and so on are available in this palette. AutoCAD provides two methods to insert the predefined hatch patterns from the **Tool Palettes**: **Drag and Drop** method and **Select and Place** method.

HATCHING AROUND TEXT, DIMENSIONS, AND ATTRIBUTES

When you select a point within a boundary to be hatched and if it contains text, dimensions, and attributes then, by default, the hatch lines do not pass through the text, dimensions, and attributes present in the object being hatched by default. AutoCAD places an imaginary box around these objects that does not allow the hatch lines to pass through it. Remember that if you are using the select objects option to select objects to hatch, you must select the text/attribute/shape along with the object in which it is placed when defining the hatch boundary. If multiple line text is to be written, the **Multiline Text** tool is used. You can also select both the boundary and the text when using the window selection method.

EDITING THE HATCH BOUNDARY

Using Grips

One of the ways you can edit the hatch boundary is by using grips. You can select the hatch pattern or hatch boundaries. If you select the hatch pattern, the hatch highlights and a grip is displayed at the centroid of the hatch. A centroid for a region that is coplanar with the *XY* plane is the center of that particular area. However, if you select an object that defines the hatch boundary, the object grips are displayed at the defining points of the selected object. Once you change the boundary definition, and if the hatch pattern is associative, AutoCAD will re-evaluate the hatch boundary and then hatch the new area.

HATCHING BLOCKS AND XREF DRAWINGS

When you apply hatching to inserted blocks and xref drawings, their internal structure is treated as if the block or xref drawing were composed of independent objects. This means that if you have inserted a block that consists of a circle within a rectangle and you want the internal circle to be hatched, you need to invoke the **Hatch** tool and then specify a point within the circle to generate the hatch, refer to Figure 12-4. However, if you choose the **Select** option from the **Boundaries** panel of the **Hatch Creation** tab, you will be prompted to select an object. Select an object; the entire block will be selected and a hatch will be created, as shown in Figure 12-4.

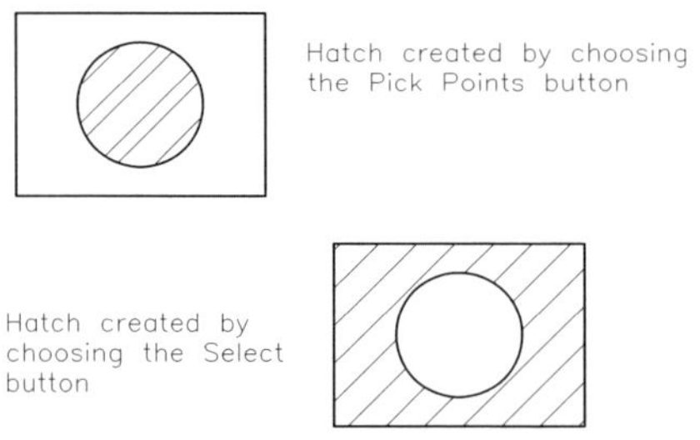

Figure 12-4 *Hatching inserted blocks*

When you xref a drawing, you can hatch any part of the drawing that is visible. Also, if you hatch an xref drawing and then use the **XCLIP** command to clip it, the hatch pattern is not clipped, although the hatch boundary associativity is removed. Similarly, when you detach the xref drawing, the hatch pattern and its boundaries are not detached, although the hatch boundary associativity is removed.

CREATING A BOUNDARY USING CLOSED LOOPS

Ribbon: Home > Draw > Hatch drop-down > Boundary
Menu Bar: Draw > Boundary **Command:** BOUNDARY/BO

The **Boundary** tool is used to create a polyline or region around a selected point within a closed area, in a manner similar to the one used for defining a hatch boundary.

Tutorial 2 Hatch

In this tutorial, you will hatch the drawing, as shown in the Figure 12-5, using various hatch patterns.

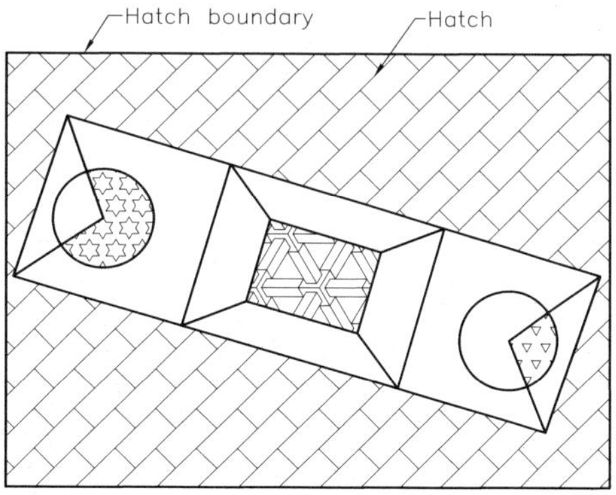

Figure 12-5 *Tutorial 2 for Hatching*

1. Make a drawing, refer to Figure 12-5.

2. Choose the **Hatch** tool from the **Draw** panel in the **Home** tab; the **Hatch Creation** tab is displayed.

3. Select the **AR-BRSTD** pattern from the **Pattern** panel and click inside the outer rectangle's boundary to select the internal pick point.

4. Next, you need to change the angle of the hatch pattern to 45°. To do so, drag the **Angle** slider in the **Properties** panel. Also, adjust the **Hatch Pattern Scale** using the **Hatch Pattern Scale** spinner.

5. Select the **Outer Island Detection** option from the **Island Detection** drop-down list in the **Options** panel, if it is not selected by default.

6. Exit the **Hatch Creation** tab by pressing ENTER.

7. Again, choose the **Hatch** tool and select the **STARS** pattern from the **Pattern** panel; you are prompted to pick an internal point in the object. Select a point in the right section of the left circle, refer to Figure 12-5; preview of the pattern is displayed. You have to specify the values of the **Hatch Angle** and **Hatch Pattern Scale** options. After making the required changes, exit the **Hatch Creation** tab.

8. Choose the **Hatch** tool and select the **ESCHER** pattern from the **Pattern** panel; you are prompted to pick an internal point in the object. Select a point in the innermost rectangle in the middle, refer to Figure 12-5; a preview will be displayed. Adjust the scale of the hatch pattern, if required and then exit.

9. Invoke the **Hatch** tool once again and select **TRIANG** from the **Pattern** panel; you are prompted to pick an internal point in the object. Select a point in the right section of the right circle, refer to Figure 12-5; preview of the pattern is displayed. Set the desired values for **Hatch Angle** and **Hatch Pattern Scale** and then exit.

EXERCISES

Exercise 1 — *Hatch Scale & Hatch Angle*

In this exercise, you will hatch the given drawing using the hatch pattern named **STEEL**. Set the scale and the angle to match the drawing shown in Figure 12-6.

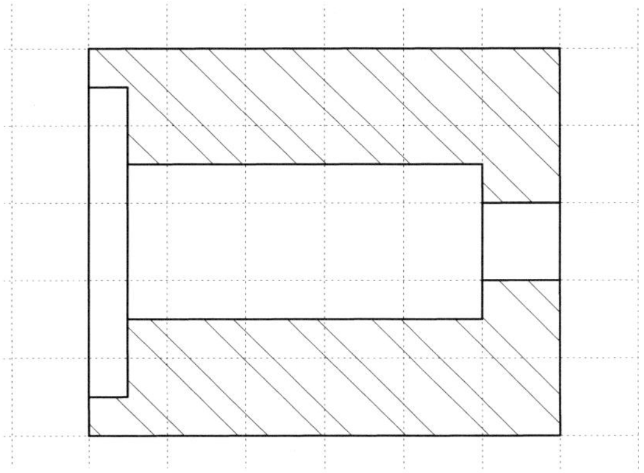

Figure 12-6 *Drawing for Exercise 1*

Exercise 2 — *Hatch Pattern*

In this exercise, you will hatch the front section view of the drawing shown in Figure 12-7 using the hatch pattern for brass. Two views, top and front are shown. In the top view, the cutting plane indicates how the section is cut and the front view shows the full section view of the object. The section lines must be drawn only where the material is actually cut.

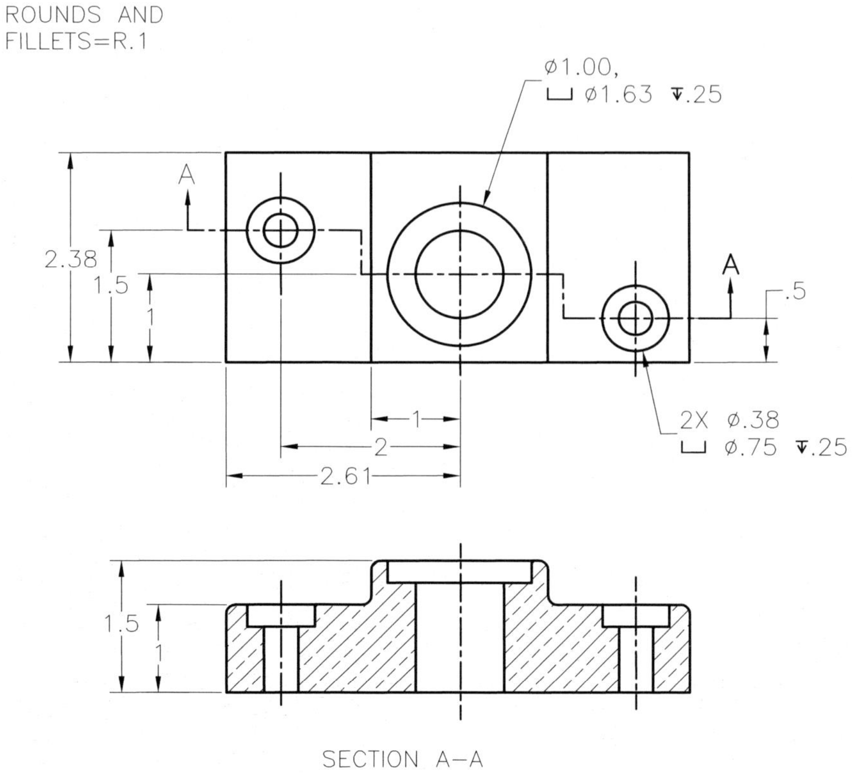

Figure 12-7 *Drawing for Exercise 2*

Chapter 13

Model Space Viewports, Paper Space Viewports, and Layouts

Learning Objectives

After completing this chapter, you will be able to:

- *Understand the concepts of model space and paper space*
- *Create tiled viewports in the model space*
- *Create floating viewports*
- *Control the visibility of viewport layers*
- *Set the linetype scaling in paper space*

MODEL SPACE AND PAPER SPACE/LAYOUTS

For ease in designing, AutoCAD provides two different types of environments, model space and paper space. The paper space is also called layout. The model space is basically used for designing or drafting work. This is the default environment that is active when you start AutoCAD. Almost the entire design is created in the model space. The paper space is used for plotting drawings or generating drawing views for the solid models.

MODEL SPACE VIEWPORTS (TILED VIEWPORTS)

Ribbon: View > Model Viewports > Viewport Configuration drop-down
Toolbar: Viewports > Display Viewports Dialog or Layouts > Display Viewports Dialog
Menu Bar: View > Viewports > New Viewports **Command:** VPORTS

A viewport in the model space is defined as a rectangular area of the drawing window in which you can create the design. When you start AutoCAD, only one viewport is displayed in the model space. You can create multiple non-overlapping viewports in the model space to display different views of the same object. Each of these viewports will act as individual drawing area. This is generally used while creating solid models. You can view the same solid model from different positions by creating the tiled viewports and defining the distinct coordinate system configuration for each viewport.

MAKING A VIEWPORT CURRENT

The viewport you are currently working in is called the current viewport. You can display several model space viewports on the screen, but you can work in only one of them at a time. You can switch from one viewport to another even when you are in the middle of a command. For example, you can specify the start point of the line in one viewport and the endpoint of the line in the other viewport. The current viewport is indicated by a border that is heavy compared to the borders of the other viewports. Also, the graphics cursor appears as a drawing cursor (screen crosshairs) only when it is within the current viewport. Outside the current viewport this cursor appears as an arrow cursor. You can enter points and select objects only from the current viewport. To make a viewport current, you can select it with the pointing device. Another method of making a viewport current is by assigning its identification number to the **CVPORT** system variable. The identification numbers of the named viewport configurations are not listed in the display.

JOINING TWO ADJACENT VIEWPORTS

Ribbon: View > Model Viewports > Join Viewports
Menu Bar: View > Viewports > Join **Command:** -VPORTS > J

AutoCAD provides you with an option of joining two adjacent viewports. However, remember that the viewports you wish to join should together form a rectangular-shaped viewport only. As mentioned earlier, the viewports in the model space can only be in rectangular shape. Therefore, you will not be able to join two viewports, in case they do not result in a rectangular shape. The viewports can be joined by using the **Join Viewports** tool available in the **Model Viewports** panel. On invoking this tool, you will be prompted to select the dominant viewport. A dominant viewport is the one whose display will be retained after joining. After selecting the dominant viewport, you will be prompted to select the viewport to be joined.

SPLITTING AND RESIZING VIEWPORTS IN MODEL SPACE

In AutoCAD, you can split the viewports in the model space. To do so enter the **-VPORTS** command; the **Enter an option [Save/Restore/Delete/Join/Single/?/2/3/4/Toggle/MOde] <3>:** prompt will be displayed. Next, enter **2**, **3**, or **4** at the command prompt to split a viewport in 2, 3, or 4 viewports, respectively.

PAPER SPACE VIEWPORTS (FLOATING VIEWPORTS)

As mentioned earlier, the viewports created in the layouts are called floating viewports. This is because unlike in model space, the viewports in the layouts can be overlapping and of any shape without any restriction. You can even convert a closed object into a viewport in the layouts. The methods of creating floating viewports are discussed next.

Creating Floating Viewports

Toolbar: Viewports > Display Viewports Dialog or Layouts > Display Viewports Dialog
Menu Bar: View > Viewports > New Viewports **Command:** VPORTS

This tool is used to create the floating viewports in layouts. However, when you invoke this tool in the layouts, the dialog box displayed is slightly modified. Instead of the **Apply to** drop-down list in the **New Viewports** tab, the **Viewport Spacing** spinner is displayed. This spinner is used to set the spacing between the adjacent viewports. The rest of the options in both the **New Viewports** and the **Named Viewports** tabs of the **Viewports** dialog box are the same as those discussed under the tiled viewports. When you select a viewport configuration and choose the **OK** button, you will be prompted to specify the first and the second corner of a box that will act as a reference for placing the viewports. You will also be given an option of **Fit**. This option fits the configuration of viewports such that they fit exactly in the current display.

Creating Rectangular Viewports

Ribbon: Layout contextual tab > Layout Viewports > Viewports drop-down > Rectangular
Menu Bar: View > Viewports > New Viewports
Toolbar: Viewports > Single Viewport **Command:** -VPORTS

To create a rectangular viewport, choose the **Rectangular** tool from **Layout > Layout Viewports > Viewports** drop-down in the **Ribbon**. The prompt sequence that will follow is given next.

Specify corner of viewport or
[ON/OFF/Fit/Shadeplot/Lock/NEw/NAmed/Object/Polygonal/Restore/LAyer/2/3/4] <Fit>: *Specify the start point of the viewport.*
Specify opposite corner: *Specify the end point of the viewport.*

Creating Polygonal Viewports

Ribbon: Layout contextual tab > Layout Viewports > Viewports drop-down > Polygonal
Menu Bar: View > Viewports > Polygonal Viewport
Toolbar: Viewports > Polygonal Viewports **Command:** -VPORTS > P

As mentioned earlier, you can create floating viewports of any closed shape. The viewports thus created can even be self-intersecting in shape. To create a polygonal viewport, choose the **Polygonal** tool from the **Layout > Layout Viewports > Viewports** drop-down.

Converting an Existing Closed Object into a Viewport

Ribbon: Layout contextual tab > Layout Viewports > Viewports drop-down > Object
Menu Bar: View > Viewports > Object **Command:** -VPORTS > O
Toolbar: Viewports > Convert Object to Viewport

This tool allows you to convert an existing closed object into a viewport. However, remember that the object selected should be a single entity. The objects that can be converted into a viewport include polygons drawn using the **Polygon** tool, rectangles drawn using the **Rectangle** tool, polylines (last segment closed using the **Close** option), circles, ellipses, closed splines, or regions.

TEMPORARY MODEL SPACE

Sometimes when you create a floating viewport in the layout, the drawing is not displayed completely inside it. In such cases, you need to zoom or pan the drawings to fit them in the viewport. But when you invoke any of the **Zoom** or **Pan** tools in the layouts, the display of the entire layout is modified instead of the display inside of the viewport. Now, to change the display of the viewports, you will have to switch to the temporary model space. The temporary model space can be invoked by choosing the **PAPER** button from the Status Bar. You can also switch to the temporary model space by double-clicking inside the viewports. The temporary model space can also be invoked using the **MSPACE** command.

Tutorial 1 — *Create Viewports*

In this tutorial, you will draw the object shown in Figure 13-1 and then create a floating viewport of the shape shown in Figure 13-2 to display the object in the layout. The dimensions of the viewport are in the paper space. Do not dimension the object.

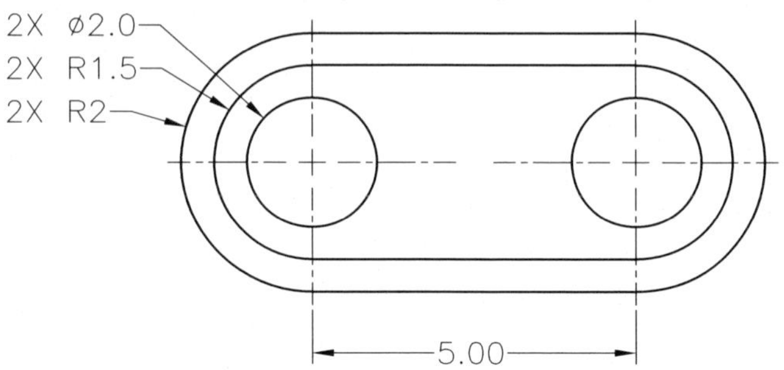

Figure 13-1 *Model for Tutorial 1*

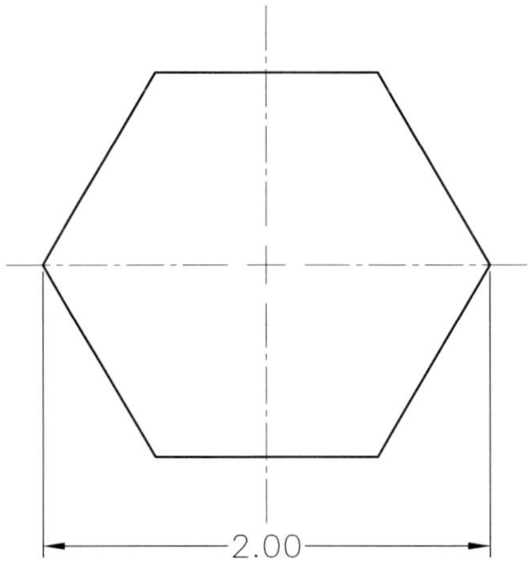

Figure 13-2 *Shape of the floating viewport*

1. Start a new drawing and then draw the object shown in Figure 13-1.

2. Choose the **Layout1** tab to switch to the layout; a rectangular viewport will be displayed in this layout.

3. Choose the **Erase** tool from the **Modify** panel of the **Home** tab; you will be prompted to select the object. Type **L** in this prompt and press ENTER to delete the last object. In this case, the last object is viewport so it will be deleted.

4. Choose the **Polygon** tool from the **Draw** panel of the **Home** tab and create the required hexagon, refer to Figure 13-2.

5. Choose the **Object** tool from **Layout > Layout Viewports > Viewports** drop-down; you will be prompted to select an object. Select the hexagon; it will be converted into a viewport. You will see that the complete object is not displayed inside the viewport. Therefore, you need to modify its display.

6. Double-click inside the viewport to switch to the temporary model space. The border of the viewport will become thick, indicating that you have switched to the temporary model space.

7. Now, using the **Zoom** and **Pan** tools, fit the drawing inside the viewport.

8. Choose the **MODEL** button from the Status Bar to switch back to the paper space. The drawing will be displayed fully inside the viewport, refer to Figure 13-3.

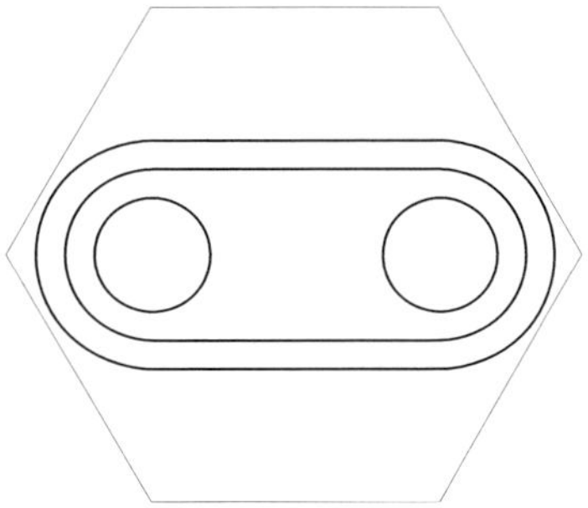

Figure 13-3 *Displaying the drawing inside the polygonal viewport*

EDITING VIEWPORTS

You can perform various editing operations on the viewports. For example, you can control the visibility of the objects in the viewports, lock their display, clip the existing viewports using an object, and so on. All these editing operations are discussed next.

Controlling the Display of Objects in Viewports

The display of the objects in the viewports can be turned on or off. If the display is turned off, the objects will not be displayed in the viewport. However, the object in the model space is not affected by this editing operation. To control the display of the objects, select the viewport entity in the layout and right-click to display the shortcut menu. In this menu, choose **Display Viewport Objects > No**.

Locking the Display of Objects in Viewports

To avoid accidental modification in the display of objects in the viewports, you can lock their display. If the display of a viewport is locked, the tools such as **Zoom** and **Pan** do not work in it. Also, you cannot modify the view in the locked viewport. To lock the display of the viewports, select it and right-click on it to display the shortcut menu. In this menu, choose **Display Locked > Yes**. Similarly, you can unlock the display of the objects in the viewports by choosing **Display Locked > No** from the shortcut menu.

Controlling the Display of Hidden Lines in Viewports

While working with three-dimensional solid or surface models, there are a number of occasions where you have to plot the solid models such that the hidden lines are not displayed. To control the display of the hidden lines, select the viewport and right-click to display the shortcut menu. In this menu, choose **Shade plot > Hidden**. Although the hidden lines will be displayed in the viewports, now they will not be displayed in the printouts.

Clipping Existing Viewports

Ribbon: Layout contextual tab > Layout Viewports > Clip
Menu Bar: Modify > Clip > Viewport
Toolbar: Viewports > Clip existing viewport **Command:** VPCLIP

You can modify the shape of any existing viewport by clipping it using an object or by defining the clipping boundary. The viewports can be clipped by choosing the **Clip** tool from the **Layout Viewports** panel in the **Layout** tab. This command can also be invoked by using the **VPCLIP** command.

INSERTING LAYOUTS

Ribbon: Layout contextual tab > Layout > Layout drop-down > New Layout
Toolbar: Layouts > New Layout **Menu Bar:** Insert > Layout > New Layout
Command: LAYOUT

The **New Layout** tool, available in the **Layout** drop-down of the **Layout** panel in the **Layout** contextual tab of the **Ribbon**, is used to create new layouts. The **LAYOUT**

command can also be used to create a new layout. This command also allows you to rename, copy, saveas, and delete existing layouts. A drawing designed in the **Model** tab can be composed for plotting in the **Layout** tab.

IMPORTING LAYOUTS TO SHEET SETS

To add a layout to the current sheet set, right-click on the **Layout** tab and then choose the **Import Layout as Sheet** option from the shortcut menu; the **Import Layout as Sheet** dialog box will be invoked. This dialog box displays all the layouts available in the selected drawing. Select the check box on the left of the drawing file name under the **Drawing Name** column and choose the **Import Checked** button to import the selected layouts into the current sheet set.

INSERTING A LAYOUT USING THE WIZARD

Command: LAYOUTWIZARD

This command displays the **Create Layout Wizard** that guides you step-by-step through the process of creating a new layout.

DEFINING PAGE SETTINGS

Ribbon: Output > Plot > Page Setup Manager **Command:** PAGESETUP
Toolbar: Layouts > Page Setup Manager

The **Page Setup Manager** tool is used to specify the layout and plot device settings for each new layout. You can also right-click on the current **Layout** tab and choose the **Page Setup Manager** option from the shortcut menu to invoke this command.

Tutorial 2

In this tutorial, you will create a drawing in the model space and then use the paper space to plot the drawing. The drawing to be plotted is shown in Figure 13-4.

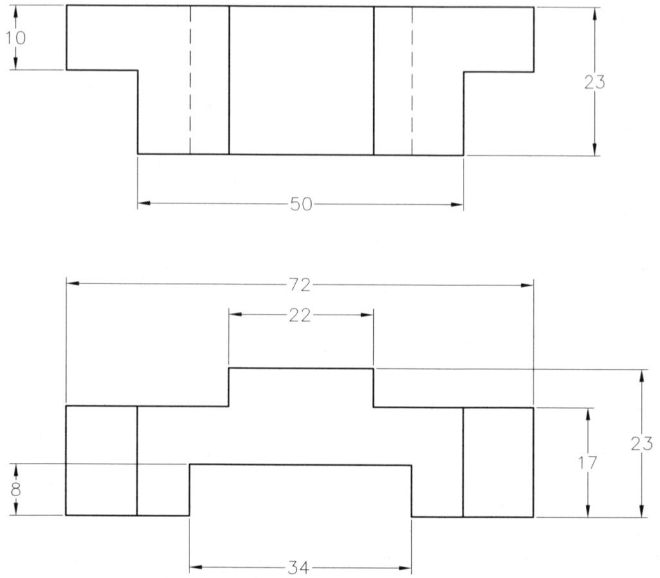

Figure 13-4 *Drawing for Tutorial 2*

1. Increase the limits to 75, 75 and then draw the sketch shown in Figure 13-4.

2. Choose the **Layout1** tab; AutoCAD displays **Layout1** with the default viewport. Delete this viewport. Right-click on the **Layout1** tab and then choose the **Page Setup Manager** option from the shortcut menu to display the **Page Setup Manager** dialog box. **Layout1** is automatically selected in the **Current page setup** list box.

3. Choose the **Modify** button to display the **Page Setup - Layout1** dialog box. Select the printer or plotter from the **Name** drop-down list in the **Printer/plotter** area. In this tutorial, **HP Lasejet4000** is used. From the drop-down list in the **Paper size** area, select the paper size that is supported by your plotting device. In this tutorial, the paper size is **ISO full bleed A4 (210.00 x 297.00 MM)**. Choose the **OK** button to accept the settings and exit the dialog box. Close the **Page Setup Manager** dialog box by choosing the **Close** button.

4. Choose the **Rectangular** tool from the **Viewports** drop-down in the **Layout Viewports** panel of the **Layout** contextual tab in the **Ribbon**; you are prompted to specify the corner point of the viewport. The prompt sequence is as follows:

   Specify corner of viewport or [ON/OFF/Fit/Shadeplot/Lock/Object/Polygonal/Restore/LAyer/2/3/4] <Fit>: **F** Enter
   Regenerating model.

5. Choose the **Extents** tool from the **Zoom** drop-down in the **Navigate** panel of the **View** tab in the **Ribbon** to zoom to the extents of the viewport.

6. Double-click in the viewport to switch to the temporary model space and repeat the previous step.

7. Create the dimension style with the text height of 1.5 and the arrowhead height of 1.25. Define all the other parameters based on the text and arrowhead heights and then select the **Annotative** check box from the **Scale for dimension features** area of the **Fit** tab in the **New Dimension Style** dialog box.

8. Using the new dimension style, dimension the drawing. Repeat step 5.

9. Double-click in the paper space to switch back to the paper space. Choose the **Plot** tool from the **Quick Access Toolbar**, the **Batch Plot** message box is displayed. Select the **Continue to plot a single sheet** option from this message box to display the **Plot - Layout1** dialog box.

10. Choose the **Window** option from the **What to plot** drop-down list in the **Plot area**; the dialog box will be closed temporarily and you will be prompted to specify the first and second corners of the window. Define a window selecting the opposite corners of the boundary of the viewport.

11. As soon as you define both the corners of the window, the **Plot** dialog box will be redisplayed on the screen. Select the **Fit to paper** check box from the **Plot scale** area.

12. Select the **Center the plot** check box from the **Plot offset (origin set to printable area)** area.

13. Choose the **Preview** button to display the plot preview. You can make any adjustments, if required, by redefining the window.

14. After you are satisfied with the preview, right-click and choose the **Plot** option from the shortcut menu. The drawing will be printed. Save this drawing with the name *Tut2.dwg*.

WORKING WITH THE MVSETUP COMMAND

The **MVSETUP** command is a very versatile command and can be used both in the **Model** tab and in the **Layout** tab. This means that this command can be used when the **Tilemode** is set to **1** or when it is set to **0**. The fuctions of this command in both the drawing environments are discussed next.

Using the MVSETUP Command in the Model Tab

On the **Model** tab, this command is used to set the units, scale factor for the drawing, and the size of the paper. On invoking this command, you will be prompted to specify whether or not you want to enable the paper space. Enter **N** at this prompt to use this command on the **Model** tab.

Using the MVSETUP Command in the Layout Tab

The way this command works in the **Layout** tab is entirely different from that in the **Model** tab. In the **Layout** tab, it is used to insert a title block, create an array of viewports, align the objects in the viewports, and so on.

EXERCISE

Exercise 1 *Tiled Viewport*

In this exercise, you will perform the following operations:

a. In the model space, make the drawing of the shaft shown in Figure 13-5.
b. Create three tiled viewports in the model space and then display the drawing in all the three tiled viewports.
c. Create a new layout with the name **Title Block** and insert a title block of ANSI A size in this layout.
d. Create two viewports, one for the drawing and one for the detail "A". Refer to Figure 13-5 for the approximate size and location. The dimensions in detail "A" viewport must not be shown in the other viewport. Also, adjust the LTSCALE factor for hidden and center lines.
e. Plot the drawing.

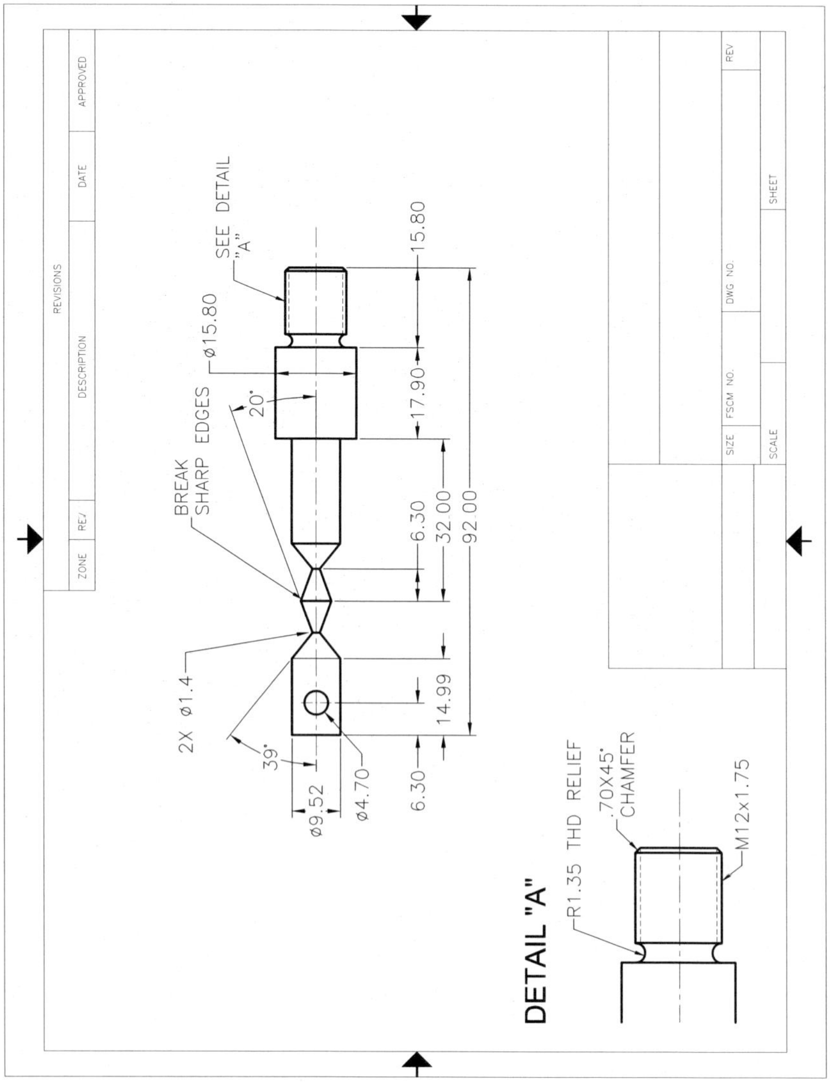

Figure 13-5 Drawing for Exercise 1

Chapter 14

Plotting Drawings

Learning Objectives

After completing this chapter, you will be able to:

- *Set plotter specifications and plot drawings*
- *Configure plotters and edit their configuration files*
- *Create, use, and modify plot styles and plot style tables*
- *Plot sheets in a sheet set*

PLOTTING DRAWINGS IN AutoCAD

After you have completed a drawing, you can store it on the computer storage device such as the hard drive or diskette. However, to get its hard copy, you should plot the drawing on a sheet of paper using a plotter or printer. A hard copy is a handy reference for professionals working on site. With pen plotters, you can obtain a high-resolution drawing. You can plot drawings in the **Model** tab or any of the layout tabs. A drawing has a **Model** and two layout tabs (**Layout1**, **Layout2**) by default. Each of these tabs has its own settings and can be used to create different plots. You can also create new layout tabs using the **New Layout** tool.

PLOTTING DRAWINGS USING THE PLOT DIALOG BOX

Ribbon: Output > Plot > Plot **Application Menu:** Print > Plot
Toolbar: Standard > Plot **Command:** Ctrl+P/PLOT
Quick Access Toolbar: Plot

The **Plot** tool is used to plot a drawing. When you choose this tool, the **Plot** dialog box is displayed. This dialog box can also be invoked by right-clicking on the **Model** tab or any of the **Layout** tabs and then choosing the **Plot** option from the shortcut menu displayed. Figure 14-1 shows the expanded **Plot** dialog box.

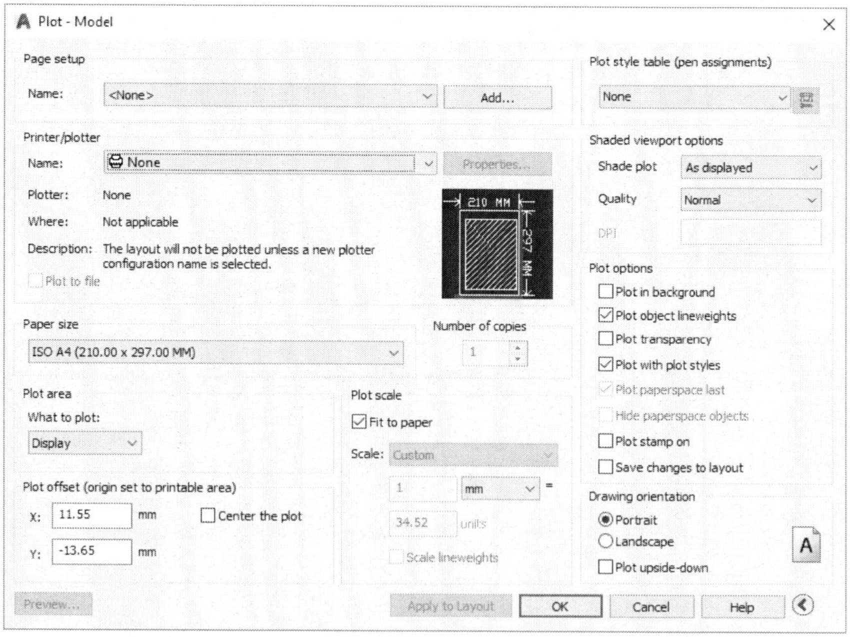

Figure 14-1 *The expanded **Plot** dialog box*

In this dialog box, some values were set when AutoCAD was first configured. You can examine these values and if they conform to your requirements, you can start plotting directly. Otherwise, you can alter these values to define plot specifications by using the options in the **Plot** dialog box.

ADDING PLOTTERS

Ribbon: Output > Plot > Plotter Manager **Application Menu:** Print > Manage Plotters
Command: PLOTTERMANAGER

When you invoke the **Plotter Manager** tool, AutoCAD will display the **Plotters** window. The **Plotters** window is basically a Windows Explorer window. It displays all the configured plotters and the **Add-A-Plotter Wizard** icon. You can right-click on any one of the icons belonging to the plotters that have already been configured to display a shortcut menu. You can choose the **Delete** option from the shortcut menu to remove a plotter from the list. You can also choose the **Rename** option from the shortcut menu to rename the plotter configuration file or choose the **Properties** option to view the properties of the configured device.

EDITING THE PLOTTER CONFIGURATION

You can modify the properties of the selected plot device by using the Plotter Configuration Editor dialog box. This dialog box can be invoked in several ways. As discussed earlier, while using the Plot or Page Setup dialog box, you can choose the Properties button in the Printer/plotter area to display the Plotter Configuration Editor dialog box. You can modify the default settings of a plotter while configuring it by choosing the Edit Plotter Configuration button on the Add Plotter - Finish page of the Add Plotter wizard. You can also select the PC3 file for editing in the Plotters window by using Windows Explorer (by default, PC3 files are stored in *C:\Users\<owner>\AppData\Roaming\Autodesk\AutoCAD 2019\R23.0\enu\Plotters)* and double-click on the file.

IMPORTING PCP/PC2 CONFIGURATION FILES

If you want to import a PCP or PC2 configuration file or plot settings created by previous releases of AutoCAD into the **Model** tab or the current layout for the drawing, you can use the **PCINWIZARD** command to display the **Import PCP or PC2 Plot Settings** wizard. All information from a PCP or PC2 file regarding plot area, plot offset, paper size, plot scale, and plot origin can be imported. Read the **Introduction** page of the wizard that is displayed carefully and then choose the **Next** button. The **Browse File name** page is displayed. Here, you can either enter the name of the PCP or PC2 file directly in the **PC2 or PCP file name** edit box or choose the **Browse** button to display the **Import** dialog box, where you can select the file to be imported. After you specify the file for importing, choose **Open** to return to the wizard. Choose the **Next** button to display the **Finish** page. After importing the files, you can modify the rest of the plot settings for the current layout.

SETTING PLOT PARAMETERS

Before starting with the drawing, you can set various plotting parameters in the **Model** tab or in the layouts tab. The plot parameters that can be set include the plotter to be used, for example, plot style table, the paper size, units, and so on. All these parameters can be set using the **Page Setup Manager** tool.

USING PLOT STYLES

The plot styles can change the complete looks of a plotted drawing. You can use this feature to override the color, linetype, and lineweight of the drawing object. For example, if an object is drawn on a layer that is assigned red color and no plot style is assigned to it, the object will be plotted as red. However, if you have assigned a plot style to the object with the color blue, the object will be plotted as blue irrespective of the layer color it was drawn on. Similarly, you can change the Linetype, Lineweight, end, join, and fill styles of the drawing, and also change the output effects such as dithering, grayscales, pen assignments, and screening. Basically, you can use Plot Styles effectively to plot the same drawing in various ways.

Every object and layer in the drawing has a plot style property. The plot style characteristics are defined in the plot style tables attached to the **Model** tab, layouts, and viewports within the layouts. You can attach and detach different plot style tables to get different looks for your plots. Generally, there are two plot style modes. They are **Color-dependent** and **Named**. The **Color-dependent** plot styles are based on object color and there are **255** color-dependent plot styles. It is possible to assign each color in the plot style a value for the different plotting properties and these settings are then saved in a color-dependent plot style table file that has a *.ctb* extension. Similarly, **Named** plot styles are independent of the object color and you can assign any plot style to any object regardless of that object's color. These settings are saved in a named plot style table file that has *.stb* extension. Every drawing in AutoCAD is in either of the plot style modes.

PLOTTING SHEETS IN A SHEET SET

Using the **Sheet Set Manager**, you can easily plot all the sheets available in a sheet set. However, before plotting the sheets in a sheet set, you need to make sure that you have selected the required printer in the page setup of all the sheets in the sheet set. This is because the printer set in the page setup of the sheet will be automatically selected to plot the sheet.

To print the sheets after setting the page setup, right-click on the name of the sheet set in the **Sheet Set Manager** and choose **Publish > Publish to Plotter** from the shortcut menu. If the value of the **BACKGROUNDPLOT** system variable is set to **2**, which is the default value, all the sheets will be automatically plotted in the background and you can continue working on the drawings.

Tutorial 1 — *Plot Style*

Create the drawing shown in Figure 14-2. Next, create a named plot style with the name *My First Table.stb* with a plot style having the following parameters:

1. Screening : 65%
2. Object line weight : 0.6000 mm
3. Line Type : ISO Dash
4. Object Color : Blue (5)

Additionally, plot the drawing using the **Date and Time** and **Paper size** stamps.

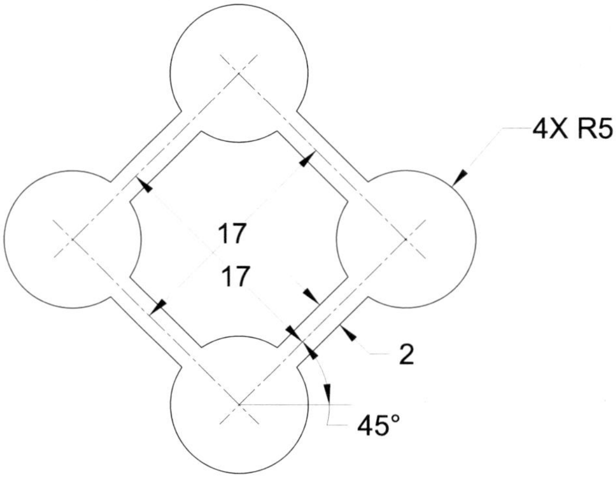

Figure 14-2 *Drawing for Tutorial 1*

1. Select the **acad-Named Plot Styles.dwt** template from **Templates** drop-down list of the **Start** tab.

2. Create the drawing, as shown in Figure 14-2. You can create the drawing by making an octagon and then converting its alternate edge into an arc with the given radius.

3. Choose the **Print > Manage Plot Styles** from the **Application Menu**; a window is displayed with all the plot styles available. Double-click on the **Add-A-Plot Style Table Wizard** shortcut icon; the **Add Plot Style Table** wizard is displayed. Choose the **Next** button; the **Begin** page is displayed.

4. Select the **Start from scratch** radio button and choose the **Next** button; the **Pick Plot Style Table** page is displayed. Now, select the **Named Plot Style Table** radio button and then choose the **Next** button; a window is displayed prompting to enter the file name. Enter **My First Table** in the **File name** text box and then choose the **Next** button, the **Add Plot Style Table - Finish** window is displayed.

5. In the **Add Plot Style Table - Finish** window, select **Plot Style Table Editor** button; a window named **Plot Style Table Editor - My First Table.stb** is displayed with the **Table View** tab chosen by default.

6. Choose the **Add Style** button at the bottom left of the window; a new column named **Style 1** will be added. Now, enter the following values in the database:

Row	**Value**
• Screening	65
• Lineweight	0.6000 mm
• Linetype	ISO Dash
• Color	Blue

Next, choose the **Save & Close** button and then choose the **Finish** button in the displayed window.

7. Now, enter the PLOTSTYLE command to invoke the **Current Plot Style** dialog box. Next, select the **My First Table.stb** plot style from the drop-down list under the **Active plot style table** area of this dialog box. Choose the **OK** button.

8. Open the **LAYER PROPERTIES MANAGER** and click under the **Plot Style** column to invoke the **Select Plot Style** dialog box. Select **Style 1** from the **Plot styles** area and choose the **OK** button.

9. Now, choose the **Plot** tool from the **Quick Access Toolbar**; the **Plot** dialog box is displayed. Next, select the **Plot stamp on** check box; the **Plot Stamp Settings** button will be displayed on the right of the check box. Choose this button; the **Plot Stamp** dialog box will be displayed. In this dialog box, select the **Date and Time** and **Paper size** check boxes. Next, clear all other check boxes. Choose the **OK** button to exit the dialog box, and then choose the **Preview** button from the **Plot-Model** dialog box to preview the plot. If the plot seems to be fine, right-click to invoke the shortcut menu. Choose the **Plot** option from the shortcut menu to take print.

EXERCISES

Exercise 1 — Plot Style

Create the drawing shown in Figure 14-3. Create a named plot style table *My Named Table.stb* with three plot styles: Style 1, Style 2, and Style 3, in addition to the Normal plot style. The Normal plot style is used for plotting the object lines. These three styles have the following specifications:

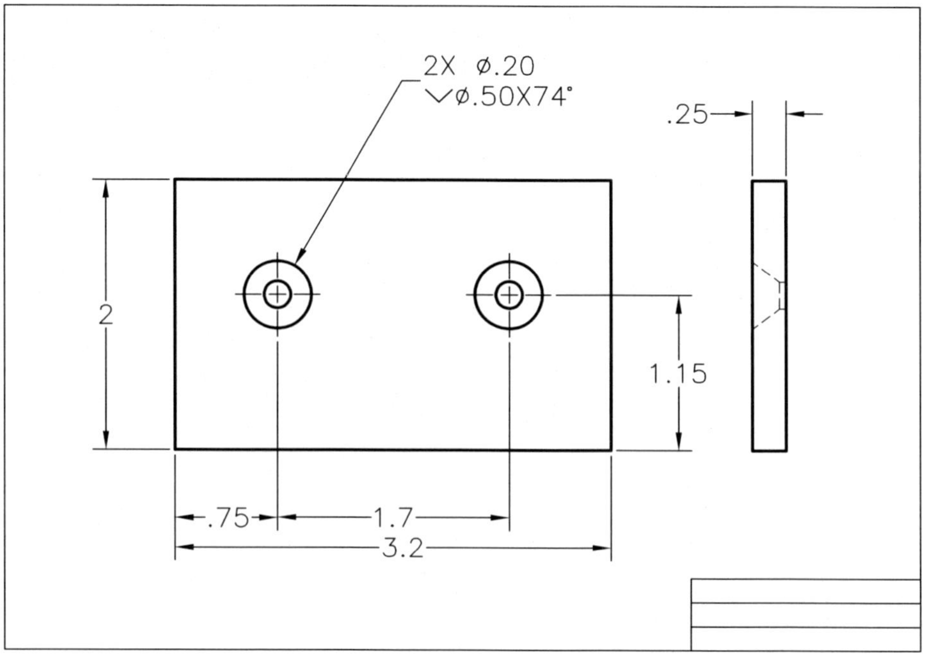

Figure 14-3 *Drawing for Exercise 1*

Style 1. This style has a value of Screening = 50. The dimensions, dimension lines, and the text in the drawing must be plotted with this style.

Style 2. This style has a value of Lineweight = 0.800. The border and title block must be plotted with this style.

Style 3. This style has a linetype of Medium Dash. The centerlines must be plotted with this plot style.

Exercise 2

Create the drawing shown in Figure 14-4 and plot it according to the following specifications. Also, create and use a plot style table with the specified plot styles.

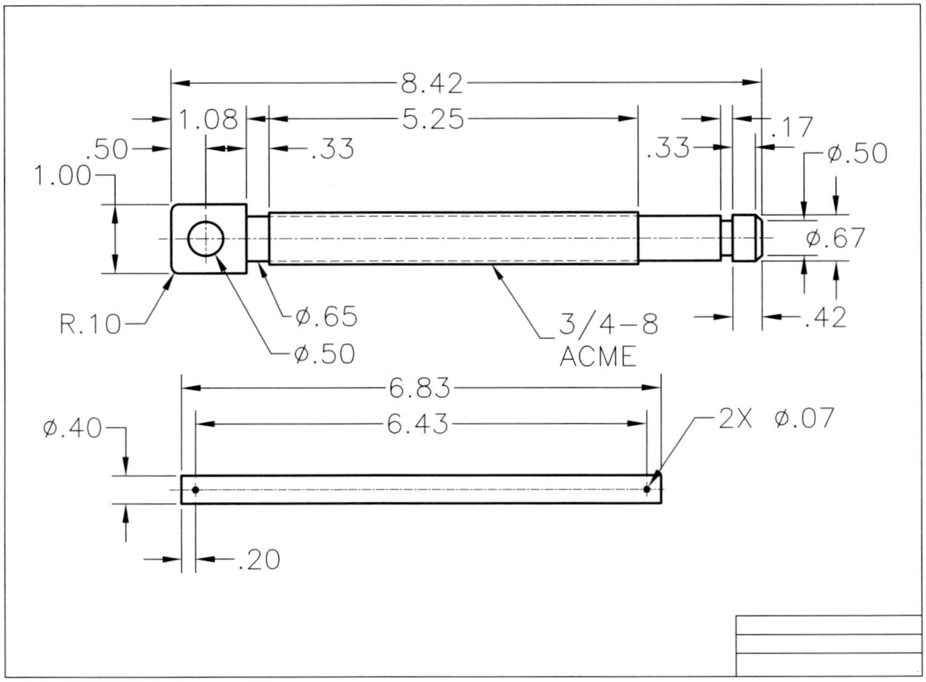

Figure 14-4 *Drawing for Exercise 2*

1. The drawing is to be plotted on 10 X 8 inch paper.
2. The object lines must be plotted with a plot style Style 1. Style 1 must have a lineweight = 0.800 mm.
3. The dimension lines must be plotted with plot style Style 2. Style 2 must have a value of screening = 50.
4. The centerlines must be plotted with plot style Style 3. Style 3 must have a linetype of Medium Dash and screening = 50.
5. The border and title block must be plotted with plot style Style 4. The value of the lineweight should be =0.25 mm.

Exercise 3

Create the drawing shown in Figure 14-5 and then plot it according to your specifications.

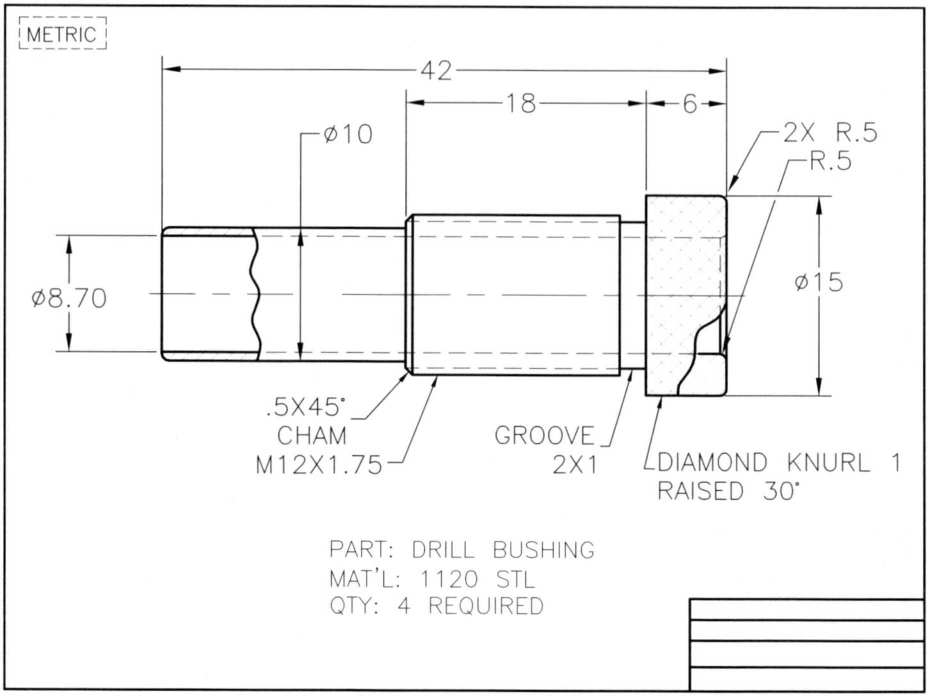

Figure 14-5 *Drawing for Exercise 3*

Chapter 15

Template Drawings

Learning Objectives

After completing this chapter, you will be able to:

- *Create template drawings*
- *Load template drawings using dialog boxes and the command line*
- *Customize drawings with layers and dimensioning specifications*
- *Customize drawings with layouts, viewports, and paper space*

CREATING TEMPLATE DRAWINGS

One way to customize AutoCAD is to create template drawings that contain initial drawing setup information and if desired, visible objects and text. When the user starts a new drawing, the settings associated with the template drawing are automatically loaded. If you start a new drawing from the scratch, AutoCAD loads default setup values. For example, the default limits are (0.0,0.0), (12.0,9.0) and the default layer is 0 with white color and a continuous linetype. Generally, these default parameters need to be reset before generating a drawing on the computer using AutoCAD. A considerable amount of time is required to set up the layers, colors, linetypes, lineweights, limits, snaps, units, text height, dimensioning variables, and other parameters. Sometimes, border lines and a title block may also be needed.

In production drawings, most of the drawing setup values remain the same. For example, the company title block, border, layers, linetypes, dimension variables, text height, LTSCALE, and other drawing setup values do not change. You will save considerable time if you save these values and reload them when starting a new drawing. You can do this by creating template drawings that contain the initial drawing setup information configured according to the company specifications. They can also contain a border, title block, tolerance table, block definitions, floating viewports in the paper space, and perhaps some notes and instructions that are common to all drawings.

STANDARD TEMPLATE DRAWINGS

AutoCAD comes with standard template drawings like *Acad.dwt*, *Acadiso.dwt*, *Acad3d.dwt*, *Acadiso3D.dwt*, *Acad-named plot styles.dwt*, *Acadiso-named plot styles.dwt*, and so on. The iso template drawings are based on the drawing standards developed by ISO (International Organization for Standardization). When you start a new drawing with **STARTUP** system variable set to 1, the **Create New Drawing** dialog box will be displayed. To load the template drawing, choose the **Use a Template** button from this dialog box; the list of standard template drawings is displayed. From this list, you can select any template drawing according to your requirements. If you want to start a drawing with the default settings, choose the **Start from Scratch** button in the **Create New Drawing** dialog box and choose the **OK** button.

Tutorial 1 — *Advance Setup Wizard*

Create a drawing template using the **Use a Wizard** button of the **Advanced Setup** wizard with the following specifications and save it with the name *proto1.dwt*.

Units	Engineering with precision 0'-0.00"
Angle	Decimal degrees with precision 0
Angle Direction	Counterclockwise
Area	144'x96'

Step 1: Setting the STARTUP system variable

Set the value of the **STARTUP** system variable to 1. Choose the **New** tool from the **Quick Access Toolbar** to display the **Create New Drawing** dialog box, as shown in Figure 15-1. Choose the **Use a Wizard** button and then select the **Advanced Setup** option. Next, choose the **OK** button; the **Units** page of the **Advanced Setup** wizard is displayed, as shown in Figure 15-2.

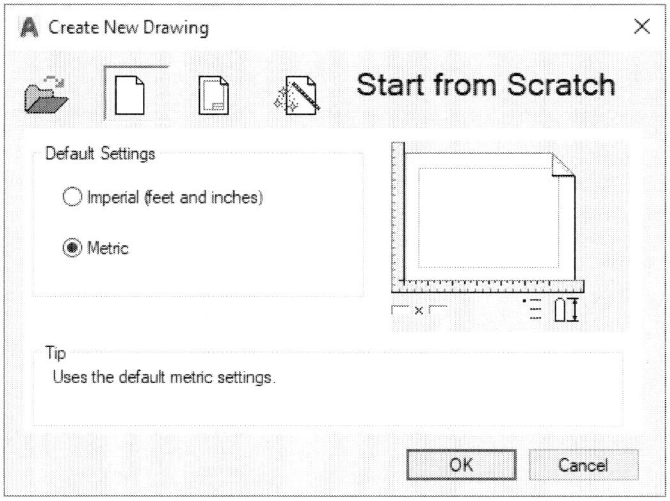

*Figure 15-1 The **Create New Drawing** dialog box*

Step 2: Setting the units of the drawing file

Select the **Engineering** radio button from the **Select the unit of measurement** area. Next, select **0'-0.00"** precision from the **Precision** drop-down list and then choose the **Next** button; the **Angle** page of the **Advanced Setup** wizard is displayed.

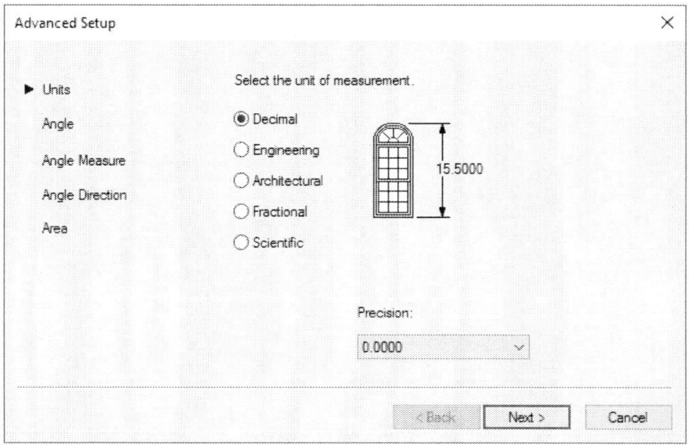

*Figure 15-2 The **Units** page of the **Advanced Setup** wizard*

Step 3: Setting the angle measurement system

In the **Angle** page, select the **Decimal Degrees** radio button, if not already selected and select **0** from the **Precision** drop-down list, as shown in Figure 15-3. Choose the **Next** button; the **Angle Measure** page of the **Advanced Setup** wizard is displayed.

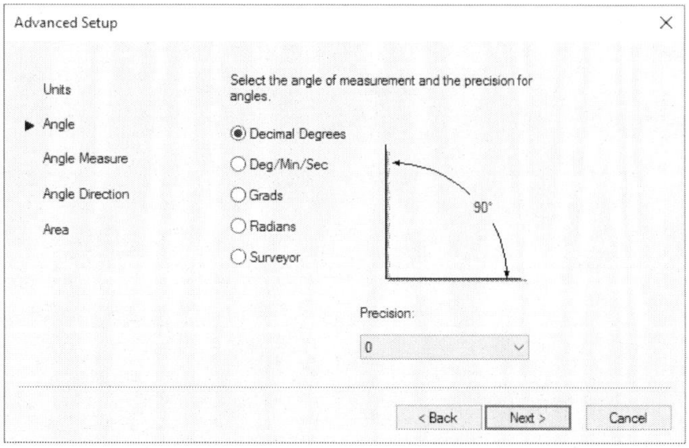

Figure 15-3 *The* ***Angle*** *page of the* ***Advanced Setup*** *wizard*

Step 4: Setting the horizontal axis for angle measurement

In the **Angle Measure** page, select the **East** radio button. Choose the **Next** button to display the **Angle Direction** page.

Step 5: Setting the angle measurement direction and drawing area

Select the **Counter-Clockwise** radio button and then choose the **Next** button; the **Area** page is displayed. Specify the area as 144' and 96' by entering the value of the width and length as **144**' and **96**' in the **Width** and **Length** edit boxes and then choose the **Finish** button. Use the **All** option from the **Zoom** drop-down list to display new limits on the screen.

Step 6: Saving the drawing as template file

Now, save the drawing as *proto1.dwt* using the **Save** tool from the **Quick Access Toolbar**. You need to select **AutoCAD Drawing Template (*.dwt)** from the **Files of type** drop-down list and enter **proto1** in the **File name** edit box in the **Save Drawing As** dialog box. Next, choose the **Save** button; the **Template Options** dialog box will be displayed on the screen, as shown in Figure 15-4. Enter the description about the template in the **Description** edit box and choose the **OK** button. Now, the drawing will be saved as *proto1.dwt* on the default drive.

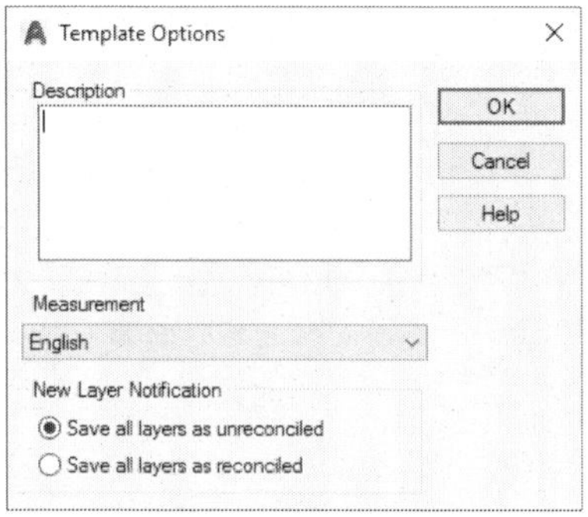

Figure 15-4 *The* ***Template Options*** *dialog box*

Note

To customize only the units and area, you can use the ***Quick Setup*** *option in the* ***Create New Drawing*** *dialog box.*

Tutorial 2 — Start from Scratch Option

Create a drawing template using the following specifications. The template should be saved with the name *proto2.dwt*.

Limits	18.0,12.0
Snap	0.25
Grid	0.50
Text height	0.125
Units	3 digits to the right of decimal point
	Decimal degrees
	2 digits to the right of decimal point
	0 angle along positive *X* axis (east)
	Angle positive if measured counterclockwise

Step 1: Starting a new drawing

Start AutoCAD and choose the **Start from Scratch** button from the **Create New Drawing** dialog box, as shown in Figure 15-5. From the **Default Settings** area, select the **Imperial (feet and inches)** radio button, refer to Figure 15-5. Choose the **OK** button to open a new file.

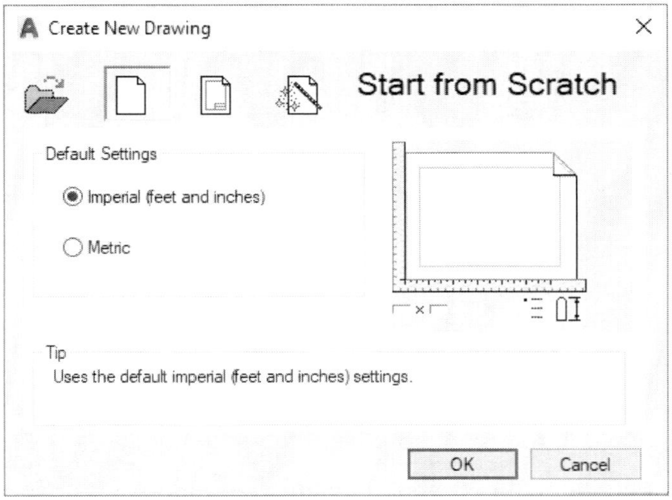

Figure 15-5 *The* ***Start from Scratch*** *button of the* ***Create New Drawing*** *dialog box*

Step 2: Setting limits, snap, grid, and text size

The **LIMITS** command can be invoked by entering **LIMITS** at the command prompt.

Command: **LIMITS**
Reset Model space limits:
Specify lower left corner or [ON/OFF] <0.0000,0.0000>: Enter
Specify upper right corner <12.0000,9.0000>: **18,12** Enter

After setting the limits, the next step is to expand the drawing display area. Select the **All** option from the **Zoom** drop-down list to display new limits on the screen.

Now, right-click on the **Grid Display** button in the Status Bar to display a shortcut menu. Choose the **Grid Settings** option from the shortcut menu to display the **Drafting Settings** dialog box. Choose the **Snap and Grid** tab. Enter **0.25** in the **Snap X spacing** and **Snap Y spacing** edit boxes in the **Snap spacing** area. Enter **0.5** in the **Grid X spacing** and **Grid Y spacing** edit boxes. Then, choose the **OK** button.

Note

*You can also use the **SNAP** and **GRID** commands to set these values.*

The size of the text can be changed by entering **TEXTSIZE** at the command prompt.

Command: **TEXTSIZE**
Enter new value for TEXTSIZE <0.2000>: **0.125** Enter

Step 3: Setting units

Choose the **Units** tool from the **Application Menu > Drawing Utilities** or enter **UNITS** at the command prompt to invoke the **Drawing Units** dialog box, as shown in Figure 15-6. In the **Length** area, select **0.000** from the **Precision** drop-down list. In the **Angle** area, select **Decimal Degrees** from the **Type** drop-down list and **0.00** from the **Precision** drop-down list. Also, make sure the **Clockwise** check box in the **Angle** area is cleared.

Choose the **Direction** button from the **Drawing Units** dialog box to display the **Direction Control** dialog box, refer to Figure 15-7 and then select the **East** radio button. Exit both the dialog boxes.

*Figure 15-6 The **Drawing Units** dialog box*

*Figure 15-7 The **Direction Control** dialog box*

Step 4: Saving the drawing as template file

Now, save the drawing as *proto2.dwt* using the **Save** tool from the **Quick Access** toolbar. You need to select **AutoCAD Drawing Template (*dwt)** from the **Files of type** drop-down list and enter **proto2** in the **File name** edit box in the **Save Drawing As** dialog box. Next, choose the **Save** button; the **Template Options** dialog box will be displayed on the screen. Enter the description about the template in the **Description** edit box and choose the **OK** button; the drawing will be saved as *proto2.dwt* on the default drive. You can also save this drawing to some other location by specifying other location from the **Save in** drop-down list of the **Save Drawing As** dialog box.

LOADING A TEMPLATE DRAWING

You can use the template drawing to start a new drawing file. To use the preset values of the template drawing, restart AutoCAD or choose the **New** tool from the **Quick Access Toolbar**. The dialog box that appears will depend on whether you have set the **STARTUP** system variable to **1** or **0**. If you have set this value as **1**, the **Create New Drawing** dialog box will appear. Choose the **Use a Template** button. All templates that are saved in the default **Template** directory will be shown in the **Select a Template** list box. If you have saved the template in any other location, choose the **Browse** button. On doing so, the **Select a template file** dialog box will be displayed. You can use this dialog box to browse the directory in which the template file is saved.

If you have set the **STARTUP** system variable to 0, then on choosing the **New** tool, the **Select template** dialog box is displayed. This dialog box also displays the default **Template** folder and all template files saved in it. You can use this dialog box to select the template file that you want to open.

CUSTOMIZING DRAWINGS WITH LAYERS AND DIMENSIONING SPECIFICATIONS

Most production drawings need multiple layers for different groups of objects. In addition to layers, it is a good practice to assign different colors to different layers to control the line width at the time of plotting. You can generate a template drawing that contains the desired number of layers with linetypes and colors according to your company specifications. You can then use this template drawing to make a new drawing. The next tutorial illustrates the procedure used for customizing a drawing with layers, linetypes, and colors.

CUSTOMIZING A DRAWING WITH LAYOUT

The Layout (paper space) provides a convenient way to plot multiple views of a three-dimensional (3D) drawing or multiple views of a regular two-dimensional (2D) drawing. It takes quite some time to set up the viewports in the model space with different vpoints and scale factors. You can create prototype drawings that contain predefined viewport settings, with vpoint and the other desired information. If you create a new drawing or insert a drawing, the views are automatically generated.

CUSTOMIZING DRAWINGS WITH VIEWPORTS

In certain applications, you may need multiple model space viewport configurations to display different views of an object. This involves setting up the desired viewports and then changing the viewpoint for different viewports. You can create a prototype drawing that contains a required number of viewports and the viewpoint information. If you insert a 3D object in one of the viewports of the prototype drawing, you will automatically get different views of the object without setting viewports or viewpoints.

CUSTOMIZING DRAWINGS ACCORDING TO PLOT SIZE AND DRAWING SCALE

For controlling the plot area, it is recommended to use layouts. You can make the drawing of any size, use the layout to specify the sheet size, and then draw the border and title block. However, you can also plot a drawing in the model space and set up the system variables so that the plotted drawing is according to your specifications. You can generate a template drawing according to plot size and scale. For example, if the scale is 1/16" = 1' and the drawing is to be plotted on a 36" by 24" area, you can calculate drawing parameters like limits, **DIMSCALE**, and **LTSCALE** and save them in a template drawing. This will save considerable time in the initial drawing setup and provide uniformity in the drawings.

EXERCISES

Exercise 1 *Relative Rectangular & Absolute Coordinates*

Create a prototype drawing (*protoe1.dwt*) with the following specifications:

Limits	36.0,24,0
Border	35.0,23.0
Grid	1.0
Snap	0.5
Text height	0.15
Units	Decimal (up to 2 places)
LTSCALE	1
Current layer	Object

LAYERS

Layer Name	Linetype	Color
0	Continuous	White
Object	Continuous	Red
Hidden	Hidden	Yellow
Center	Center	Green
Dim	Continuous	Blue
Border	Continuous	Magenta
Notes	Continuous	White

This prototype drawing should have a border line and a title block, as shown in Figure 15-8.

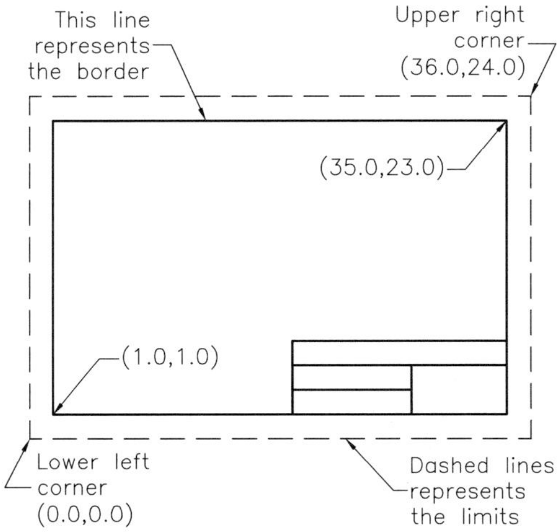

Figure 15-8 *Prototype drawing*

Exercise 2 *Template with Plot Sheet Size & Title Block*

Create a template drawing shown in Figure 15-9 with the following specifications and save it with the name *protoe2.dwt*:

Plotted sheet size	36" x 24" (Figure 15-9)
Scale	1/2" = 1.0'
Text height	1/4" on plotted drawing
LTSCALE	24
DIMSCALE	24
Units	Architectural
	32-denominator of smallest fraction to display
	Angle in degrees/minutes/seconds
	Precision 0d00"00"
	Angle positive if measured counterclockwise
Border	Border is 1-1/2" inside the edges of the plotted drawing sheet, using the PLINE 1/32" wide when plotted.

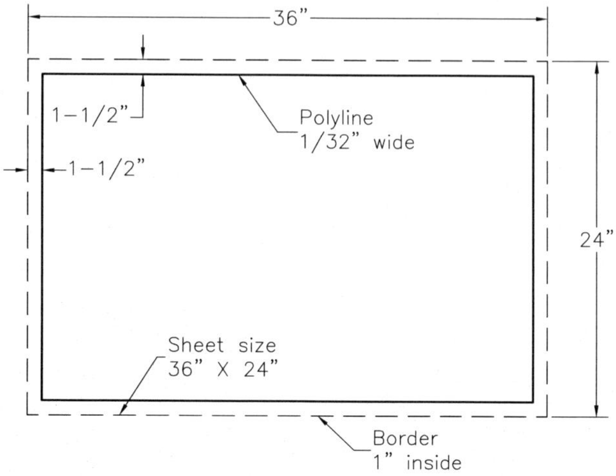

Figure 15-9 *Drawing for Exercise 2*

Chapter 16

Working with Blocks

Learning Objectives

After completing this chapter, you will be able to:

- *Create and insert blocks*
- *Add parameters and assign actions to the blocks to make them dynamic blocks*
- *Create drawing files by using the Write Block dialog box*
- *Use the DesignCenter to locate, preview, copy, or insert blocks and existing drawings*
- *Use the Tool Palettes to insert blocks*
- *Edit blocks*
- *Rename blocks and delete unused blocks*

THE CONCEPT OF BLOCKS

The ability to store parts of a drawing, or the entire drawing, such that they need not be redrawn when required in the same drawing or another drawing is a great benefit to the user. These parts of a drawing, entire drawings, or symbols (also known as blocks) can be placed (inserted) in a drawing at the location of your choice, with the desired orientation, and scale factor. A block is given a name (block name) and is referenced (inserted) by its name. All objects within a block are treated as a single object. You can move, erase, or list the block as a single object, that is, you can select the entire block simply by clicking anywhere on it. As for the edit and inquiry commands, the internal structure of a block is immaterial, since a block is treated as a primitive object, like a polygon. If a block definition is changed, all references to the block in the drawing are updated to incorporate the changes.

CONVERTING ENTITIES INTO A BLOCK

Ribbon: Insert > Block Definition > Block drop-down > Create Block or Home > Block > Create Block
Toolbar: Draw > Make Block **Command:** BLOCK/B

You can convert the entities in the drawing window into a block by using the **BLOCK** command or by choosing the **Create Block** tool from the **Block** panel of the **Home** tab. Alternatively, you can do so by choosing the **Make Block** tool from the **Draw** toolbar. When you invoke this tool, the **Block Definition** dialog box will be displayed. You can use the **Block Definition** dialog box to save any objects as a block. In the **Name** edit box of the **Block Definition** dialog box, enter the name of the block you want to create.

INSERTING BLOCKS

Ribbon: Home/Insert > Block > Insert Block **Command:** INSERT/I
Toolbar: Insert > Insert Block or Draw > Insert drop-down > Insert Block

The blocks created in the current drawing are inserted using the **Insert Block** tool. An inserted block is called a block reference. To insert a block, choose the **Insert Block** tool; a flyout will be displayed. In the flyout, the blocks available in the drawing are displayed. Click on the required block; the selected block will be attached to the cursor. Specify the required location to insert the block; the block will be inserted in the drawing.

CREATING AND INSERTING ANNOTATIVE BLOCKS

You can create an annotative block by selecting the **Annotative** check box from the **Behavior** area of the **Block Definition** dialog box. The annotative block acquires the annotative properties like text, dimension, hatch, and so on. To convert an existing non-annotative block into an annotative one, choose the **Create Block** tool from the **Create Block** drop-down from the **Block Definition** panel of the **Insert** tab; the **Block Definition** dialog box will be displayed. Select the block to be converted to annotative from the **Name** drop-down list of the **Block Definition** dialog box. Select the **Annotative** check box from the **Behavior** area and then choose the **OK** button. AutoCAD will display a message box stating that the block definition has changed. Do you want to redefine it? If you want to redefine the block, choose the **Redefine block** button; the specified block will be converted to annotative.

Tutorial 1 Annotative Block

In this tutorial, you will draw the object shown in Figure 16-2, and convert it into an annotative block, named NOR Gate. Next, you will insert the NOR Gate block into the drawing at the annotation scales of 1:1, 1:2, and 1:8 and notice the changes in the size of the annotative blocks inserted in the drawing.

1. Start a new file in the **Drafting & Annotation** workspace and draw the object, refer to Figure 16-1.

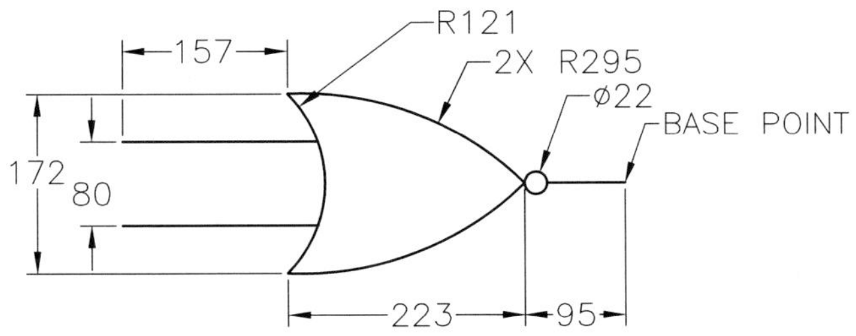

Figure 16-1 *Drawing of block for Tutorial 1*

2. Invoke the **Block Definition** dialog box by choosing the **Create Block** tool from the **Create Block** drop-down of the **Block Definition** panel in the **Insert** tab.

3. Enter **NOR Gate** as the name of the block in the **Name** edit box. Next, choose the **Select Objects** button from the **Objects** area; the **Block Definition** dialog box disappears. Select the object drawn and press the ENTER key; the **Block Definition** dialog box appears. Next, select the **Delete** radio button from the **Objects** area.

4. Choose the **Pick point** button from the **Base point** area; the **Block Definition** dialog box disappears from the screen. Specify the base point, as shown in Figure 16-1.

5. Select the **Annotative** check box from the **Behavior** area. Next, choose the **OK** button from the **Block Definition** dialog box; the selected objects disappear from the screen and an annotative block is defined with the name **NOR Gate**.

Before proceeding further, ensure that the **Add scales to annotative objects when the annotation scale changes** button is chosen in the Status Bar. Now, you need to change the annotation scale of the drawing to 1:8.

6. Click on the down-arrow on the right of the **Annotation Scale** button in the Status Bar, and choose the scale **1:8** from the flyout displayed.

7. To insert the block into the drawing at the annotation scale of 1:8, choose the **Insert Block** tool from the **Block** panel in the **Insert** tab and then choose **More Options** from the drop-down; the **Insert** dialog box is displayed.

8. Select the **NOR Gate** block from the **Name** drop-down list. Select the **Specify On-screen** check box from the **Insertion point** area and choose the **OK** button.

9. Specify the insertion point for the block by clicking on the screen; the block is inserted into the drawing at an annotation scale of 1:8.

10. Similarly, set the annotation scale to **1:2** in the Status Bar and insert the block. Next, set the annotation scale to **1:1** and insert the block.

11. Turn off the **Add scales to annotative objects when the annotation scale changes** button from the status bar. Next, change the scale to 1:8, 1:2, and 1:1 and you will notice the difference in the sizes of the inserted blocks, refer to Figure 16-2. This automated variation in the size occurs due to annotative blocks.

12. Select the block to which you had added the annotation scales; the preview of the block with all annotation scales associated to the block is displayed, as shown in Figure 16-3. The bigger block with the current annotation (1:8) scale and the smaller blocks with other annotation scales (1:1, 1:2) are displayed with the faded dashed lines.

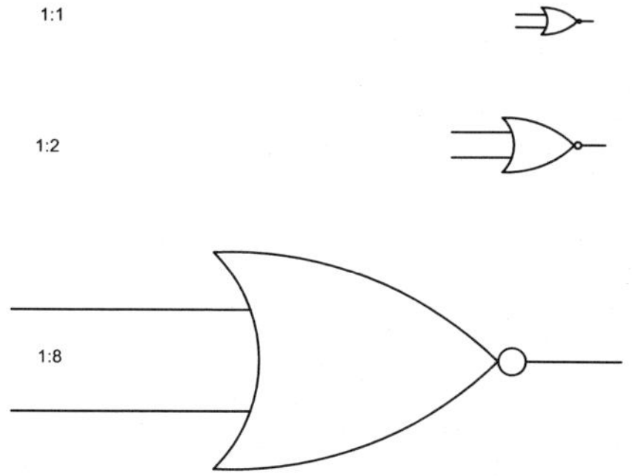

Figure 16-2 *Annotative blocks inserted at different annotation scales*

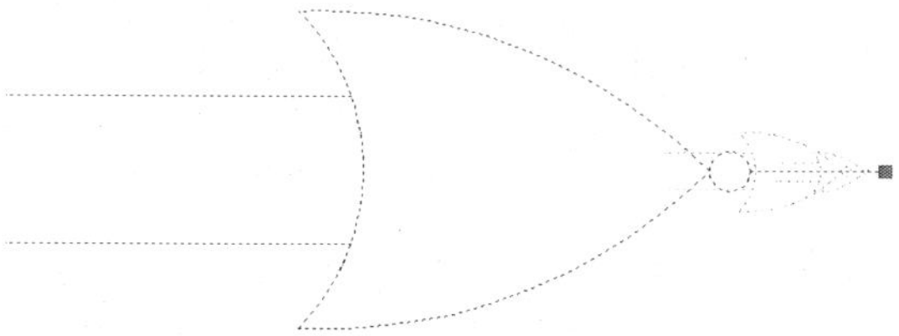

Figure 16-3 *Different annotation scales associated to a single block*

13. To move the block with the current annotation scale (1:8) to the other locations, select the blue grip displayed at the insertion point and move the block to the desired location. In this way, you can place the block with different annotation scales to any desired location.

Block Editor

Ribbon: Insert > Block Definition > Block Editor or Home > Block > Block Editor
Toolbar: Standard > Block Editor **Command:** BEDIT/BE

The **Block Editor** tool is used to edit existing blocks or create new blocks. To invoke the **Block Editor**, choose the **Block Editor** tool from the **Block Definition** panel of the **Insert** tab or enter **BE** (shortcut for the **BEDIT** command) at the command line. You can also double-click on the existing block to edit the block. On doing so, the **Edit Block Definition** dialog box will be displayed.

If you want to create a new block, enter its name in the **Block to create or edit** text box and choose the **OK** button; the **Block Editor** will be invoked in which you can draw the entities in the new block. Similarly, if you want to edit a block, select it from the list box provided in this dialog box; its preview will be displayed in the **Preview** area. Also, its related description, if any, will be displayed in the **Description** area. Choose the **OK** button; the **Block Editor** will be invoked.

DYNAMIC BLOCKS

Dynamic blocks provides you the flexibility of modifying the geometry of the inserted blocks dynamically or using the **PROPERTIES** palette. You can create dynamic blocks by adding Parameters and Actions to existing blocks using the **Action Parameters** panel in the **Block Editor** contextual tab or the **BLOCK AUTHORING PALETTES**.

Block Editor Tab

The **Block Editor** contextual tab, provides the tools to create and modify dynamic blocks. Additionally, you can create visibility states for dynamic blocks using the **Visibility** panel of the **Block Editor** contextual tab.

Adding Parameters and Assigning Actions to Dynamic Blocks (BLOCK AUTHORING PALETTES)

The **BLOCK AUTHORING PALETTES** contains the tools to add parameters and actions to the dynamic blocks. You can also invoke these tools from the **Parameters** drop-down in the **Action Parameters** panel of the **Block Editor** contextual tab.

INSERTING BLOCKS USING THE DESIGNCENTER

You can use the **DESIGNCENTER** to locate, preview, copy, or insert blocks from existing drawings into the current drawing. To insert a block, choose the **DesignCenter** tool from the **Palettes** panel in the **View** tab to display the **DESIGNCENTER** palette. By default, the tree pane is displayed. Choose the **Tree View Toggle** button to display the tree pane on the left side, if it is not already displayed. Expand **Computer** to display *C:/Program Files/Autodesk/AutoCAD 2019/Sample/en-us/DesignCenter* folder by clicking on the plus (+) signs on the left of the respective folders. Click on the plus sign adjacent to the folder to display its contents. Select a drawing file you wish to use to insert blocks from and then click on the plus sign adjacent to the drawing again. All the icons depicting the components such as blocks, dimension styles, layers, linetypes, text styles, and so on, in the selected drawing are displayed. Select **Blocks** by clicking on it in the tree pane; the blocks in the drawing are displayed in the palette. Select the block you wish to insert and drag and drop it into the current drawing. Later, you can move it in the drawing to the desired location.

USING TOOL PALETTES TO INSERT BLOCKS

You can use the **TOOL PALETTES** to insert predefined blocks in the current drawing. The **TOOL PALETTES** has many tabs. You can view the complete list of tabs by clicking on stacks of the tabs available at the end of the list of tabs. In this chapter, you will learn how to insert blocks using the tabs of the **TOOL PALETTES**.

Inserting Blocks in the Drawing

AutoCAD provides two methods to insert blocks from the **TOOL PALETTES**: Drag and Drop method and Select and Place method. Both methods of inserting blocks using the **TOOL PALETTES** are discussed next.

Drag and Drop Method

To insert blocks from the **TOOL PALETTES** in the drawing using this method, move the cursor over the desired predefined block in the **TOOL PALETTES**. You will notice that as you move the cursor over the block, a tooltip is displayed that shows the name and description of the block. Press and hold the left mouse button and drag the cursor to the drawing area. Release the left mouse button and you will notice that the selected block is inserted in the drawing.

Select and Place Method

You can also insert the desired block in the drawings using the select and place method. To insert the block using this method, move the cursor over the desired block in the **TOOL PALETTES**; a tooltip is displayed that shows the name and description of the block. Press the left mouse button; the selected block is attached to the cursor and the **Specify insertion point or [Basepoint/Scale/X/Y/Z/Rotate]** prompt is displayed. Modify any parameter using the prompt sequence and then move the cursor to the required location in the drawing area. Click the left mouse button; the selected block is inserted at the specified location.

ADDING BLOCKS IN TOOL PALETTES

By default, the **TOOL PALETTES** displays the predefined blocks in AutoCAD. You can also add the desired block and the drawing file to the **TOOL PALETTES**. This is done using the **DESIGNCENTER**. AutoCAD provides two methods for adding blocks from the **DESIGNCENTER** to the **TOOL PALETTES**; Drag and Drop method and shortcut menu. These two methods are discussed next.

Drag and Drop Method

To add blocks from the **DESIGNCENTER** in the **TOOL PALETTES**, move the cursor over the desired block in the **DESIGNCENTER**. Press and hold the left mouse button on the block and drag the cursor to the **TOOL PALETTES**.

Shortcut Menu

You can also add the desired block from the **DESIGNCENTER** to the **TOOL PALETTES** using the shortcut menu. To add the block, move the cursor over the desired block in the **DESIGNCENTER** and right-click on it to display a shortcut menu. Choose **Create Tool Palette** from it. You will notice that a new tab with the name **New Palette** is added to the **TOOL PALETTES**. And, the block is added in the new tab of the **TOOL PALETTES**.

MODIFYING EXISTING BLOCKS IN THE TOOL PALETTES

If you modify an existing block that was added to the **TOOL PALETTES** and then insert it using the **TOOL PALETTES** in the same or a new drawing, you will notice that the modified block is inserted and not the original block. However, if you insert the modified block from the **TOOL PALETTES** in the drawing in which the original block was already inserted, AutoCAD inserts the original block and not the modified one. This is because the file already has a block of the same name in its memory.

To insert the modified block, you first need to delete the original block from the current drawing using the **Erase** tool. Next, you need to delete the block from the memory of the current drawing. The unused block can be deleted from the memory of the current drawing using the **PURGE** command. To invoke this command, enter **PURGE** at the command prompt. The **Purge** dialog box is displayed. Choose the (+) sign located on the left of **Blocks** in the tree view available in the **Items not used in drawing** area. You will notice that a list of blocks in the drawing is shown. Select the original block to be deleted from the memory of the current drawing and then choose the **Purge** button. The **Confirm Purge** dialog box is displayed, which confirms the purging of the selected item. Choose **Purge this item** and then choose the **Close** button to exit the **Purge** dialog box. Next, when you insert the block using the **TOOL PALETTES**, the modified block is inserted in the drawing.

NESTING OF BLOCKS

The concept of having one block within another block is known as the nesting of blocks. For example, you can insert several blocks by selecting them, and then with the **Create Block** tool, create another block. Similarly, if you use the **Insert** tool to insert a drawing containing several blocks into the current drawing, it creates a block containing nested blocks in the current drawing. There is no limit to the degree of nesting. The only limitation in nesting of blocks is that blocks that reference themselves cannot be inserted. The nested blocks must have different block names. Nesting of blocks affects layers, colors, and linetypes.

Tutorial 2 — Nested Block

1. Change the color of layer 0 to red.
2. Draw a circle with color **By Block** and then form its block, B1. It appears white because its color is set to **By Block**, refer to Figure 16-4.
3. Set the color to **By Layer** and draw a square. The color of the square is red.
4. Insert block B1. Notice that the block B1 (circle) assumes red color.
5. Create another block B2 consisting of the Block B1 (circle) and square.
6. Create a layer L1 with green color and hidden linetype. Make it current. Insert block B2 in layer L1.
7. Explode block B2. Notice the change.
8. Explode block B1, circle. You will notice that the color of the circle changes to white because it was drawn with the color set to **By Block**.

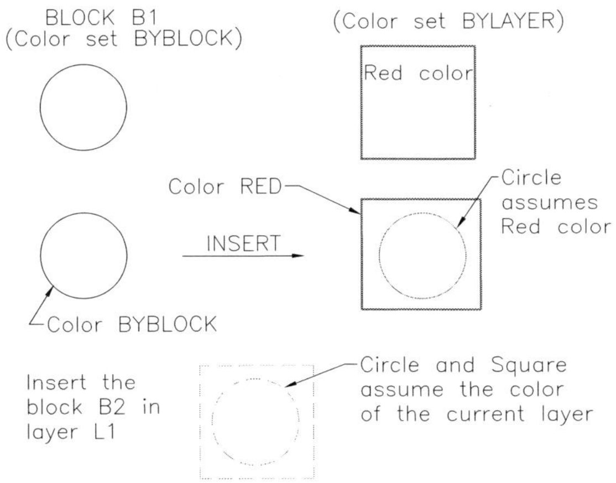

Figure 16-4 *Blocks with different layers and colors*

INSERTING MULTIPLE BLOCKS

Command: MINSERT

The **MINSERT** (multiple insert) command is used for the multiple insertion of a block. Also, blocks inserted by using **MINSERT** cannot be exploded. With the **MINSERT** command, only one block reference is created. But, in addition to the standard features of a block definition (insert point, X/Y scaling, rotation angle, and so on), this block has repeated row and column counts. In this manner, using this command saves time and disk space. The prompt sequence is very similar to that of the **-INSERT** and **-ARRAY** commands.

CREATING DRAWING FILES USING THE WRITE BLOCK DIALOG BOX

Ribbon: Insert > Block Definition > Create Block drop-down > Write Block
Command: WBLOCK/W

The blocks are symbols created by the **Create Block** tool and can be used only in the drawing in which they were created. This is a shortcoming because you may need to use a particular block in different drawings. The **Write Block** tool is used to export symbols by writing them to new drawing files that can be inserted in any drawing. With the **Write Block** tool, you can create a drawing file (*.dwg* extension) of the specified blocks, selected objects in the current drawing, or the entire drawing. All the used named objects (linetypes, layers, styles, and system variables) of the current drawing are inherited by the new drawing created with the **Write Block** tool. This block can then be inserted in any drawing.

When you invoke the **Write Block** tool, the **Write Block** dialog box is displayed. This dialog box converts the blocks into drawing files and also saves objects as drawing files. You can also save the entire current drawing as a new drawing file.

DEFINING THE INSERTION BASE POINT

Ribbon: Insert > Block Definition > Set Base Point or Home > Block > Set Base Point
Command: BASE

The **Set Base Point** tool lets you set the insertion base point for a drawing, just as you set the base insertion point using the **Create Block** tool. This base point is defined when you insert a drawing into some other drawings, the specified base point will be placed on the insertion point. By default, the base point is at the origin (0,0,0). When a drawing is inserted on a current layer, it does not inherit the color, linetype, or thickness properties of the current layer.

EDITING BLOCKS

Ribbon: Insert > Reference > Edit Reference **Command:** REFEDIT
Toolbar: Refedit > Edit Reference In-Place

You can edit blocks by breaking them into parts and then making modifications and redefining them or by editing them in place. Both these methods have been discussed here.

Editing Blocks in Place

You can edit blocks in the current drawing by using the **Edit Reference** tool, referred to as the in-place reference editing. This feature of AutoCAD allows you to make minor changes to blocks, wblocks, or drawings that have been inserted in the current drawing without breaking them up into component parts or opening the original drawing and redefining them. This command saves valuable time of going back and forth between drawings and redefining blocks.

Exploding Blocks Using the XPLODE Command

Command: XPLODE

With the **XPLODE** command, you can explode a block or blocks into component objects and simultaneously control their properties such as layer, linetype, color, and lineweight. The scale factor of the object to be exploded should be equal. Note that if the scale factor of the objects to be exploded is not equal, you need to change the value of the **EXPLMODE** system variable to 1. Note that if the **Allow exploding** check box in the **Behavior** area of the **Block Definition** dialog box was cleared while creating the block, you will not be able to explode the block.

RENAMING BLOCKS

Command: RENAME

Blocks can be renamed using the **RENAME** command. To rename a block, enter **RENAME** at the command prompt and press ENTER; the **Rename** dialog box will be displayed. In this dialog box, the **Named Objects** list box displays the categories of object types that can be renamed, such as blocks, layers, dimension styles, linetypes, Multileader styles, material, table styles and text styles, UCSs, views, and viewports. You can rename all of these objects except layer 0 and continuous linetype. When you select **Blocks** from the **Named Objects** list, the **Items** list box displays all the block names in the current drawing. When you select a block name to rename from the **Items** list box, it is displayed in the **Old Name** edit box. Enter the new name to be assigned to the block in the **Rename To** edit box. Choosing the **Rename To** button applies the change in name to the old name. Choose the **OK** button to exit the dialog box.

DELETING UNUSED BLOCKS

Sometimes after completing a drawing, you may notice that the drawing contains several named objects, such as dimstyles, textstyles, layers, blocks, and so on that are not being used. Since these unused named objects unnecessarily occupy disk space, you may want to remove them. Unused blocks can be deleted with the **-PURGE** command.

EXERCISES

Exercise 1 *Block*

Draw a circle of 1 unit radius and then draw multiple circles inside it, as shown in Figure 16-5. Next, convert them into a block and name the entity as CIRCLE. Refer to the following figure for details.

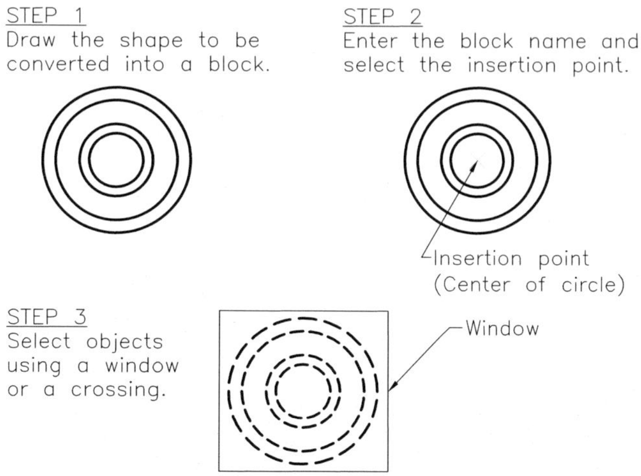

Figure 16-5 *Block created from multiple circles*

Exercise 2 Constraints to Block

Create the sketch of a bolt, as shown in Figure 16-6. Next, convert it to a block and apply constraints such that you can create bolts of diameter 6, 10, 12.5, 15, 18, 24, and 25.

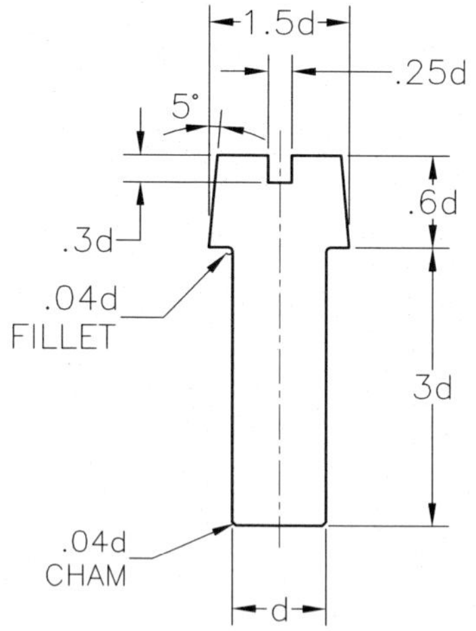

Figure 16-6 *Drawing for Exercise 2*

Chapter 17

Defining Block Attributes

Learning Objectives

After completing this chapter, you will be able to:

- *Understand block attributes*
- *Create annotative block attributes*
- *Edit attribute tag names*
- *Insert blocks with attributes and assign values to attributes*
- *Extract attribute values from inserted blocks*

UNDERSTANDING ATTRIBUTES

AutoCAD has provided a facility that allows the user to attach information to blocks. This information can then be retrieved and processed by other programs for various purposes. For example, you can use this information to create a bill of material for a project, find total number of computers in a building, or determine the location of each block in a drawing. Attributes can also be used to create blocks (such as title blocks) with prompted or preformatted text to control text placement. The information associated with a block is known as attribute value or attribute. AutoCAD references the attributes with a block through tag names. Before assigning attributes to a block, you must create an attribute definition by using the **Define Attributes** tool. The attribute definition describes the characteristics of the attribute.

DEFINING ATTRIBUTES

Ribbon: Insert > Block Definition > Define Attributes **Command:** ATTDEF/ATT

When you invoke the **Define Attributes** tool, the **Attribute Definition** dialog box will be displayed, as shown in Figure 17-1. The block attributes can be defined using this dialog box. When creating an attribute definition, you must define the Mode, Attribute, Insertion Point, and Text Settings for each attribute. All these information can be entered in the dialog box.

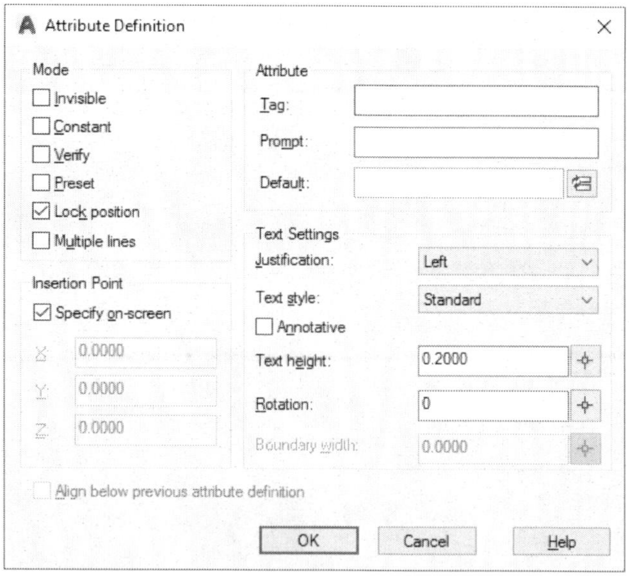

Figure 17-1 *The* ***Attribute Definition*** *dialog box*

Tutorial 1 — Attribute Definition

In this tutorial, you will define the following attributes for a computer and then create a block using the **Create Block** tool. The name of the block is COMP.

Mode	**Tag name**	**Prompt**	**Default value**
Constant	ITEM		Computer
Preset, Verify	MAKE	Enter make:	CAD-CIM
Verify	PROCESSOR	Enter processor type:	Unknown
Verify	HD	Enter Hard-Drive size:	40 GB
Invisible, Verify	RAM	Enter RAM:	256 MB

1. Draw the sketch of a computer and convert it into a block, as shown in Figure 17-2. Assume the dimensions, or measure the dimensions of the computer you are using for AutoCAD.

2. Choose the **Define Attributes** tool from the **Block Definition** panel in the **Insert** tab; the **Attribute Definition** dialog box is displayed.

3. Define the first attribute shown in the preceding table. Select the **Constant** check box in the **Mode** area. In the **Tag** edit box, enter the tag name, **ITEM**. Similarly, enter **COMPUTER** in the **Default** edit box. Note that the **Prompt** edit box is not available because the mode is constant.

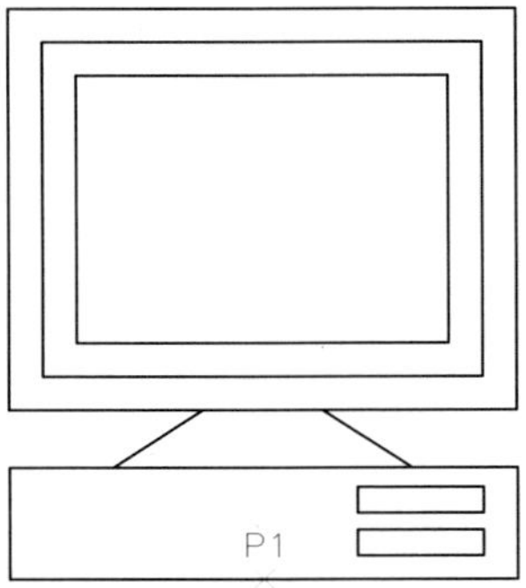

Figure 17-2 *Drawing for Tutorial 1*

4. In the **Insertion Point** area, select the **Specify on-screen** check box to define the text insertion point, if not selected by default.

5. In the **Text Settings** area, specify the justification, style, annotative property, height, and rotation of the text.

6. Choose the **OK** button once you have entered information in the **Attribute Definition** dialog box. Select a point below the insertion base point (P1) of the computer to place the text.

7. Press ENTER to invoke the **Attribute Definition** dialog box again. Enter the mode and attribute information for the second attribute as shown in the table at the beginning of Tutorial 1. You need not define the insertion point and text options again. Select the **Align below previous attribute definition** check box that is located just below the **Insertion Point** area. Now, choose the **OK** button. AutoCAD places the attribute text just below the previous attribute text.

8. Similarly, define the remaining attributes also, refer to Figure 17-3.

9. Now, use the **Create Block** tool to create a block. Make sure that the **Retain** radio button is selected in the **Objects** area. The name of the block is **COMP**, and the insertion point of the block is P1, midpoint of the base. When you select objects for the block, make sure you also select attributes.

10. Insert the block created. You will notice that the order of prompts is the same as the order of attributes selection.

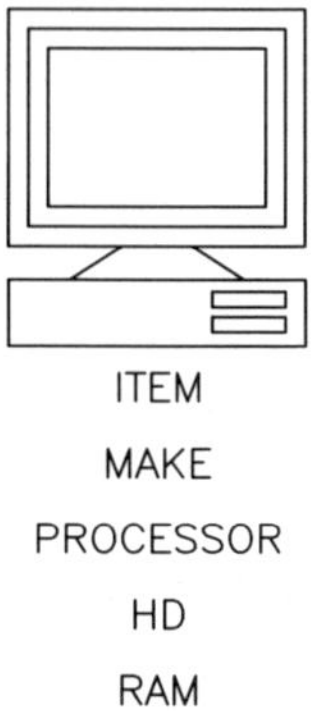

Figure 17-3 *Attributes defined below the computer drawing*

EDITING ATTRIBUTE DEFINITION

Menu Bar: Modify > Object > Text > Edit **Command:** TEXTEDIT

You can edit the text and attribute definitions before you define a block, using the **Edit** tool from the **Modify > Object > Text** menu. After invoking this tool, AutoCAD will prompt you to **Select an annotation object or [Undo/Mode]**. If you select an attribute, then the **Edit Attribute Definition** dialog box is displayed.

You can also invoke the **Edit Attribute Definition** dialog box by double-clicking on the attribute definition. You can enter the new values in the respective edit boxes. Once you have entered the changed values, choose the **OK** button in the dialog box. If you have selected the **Multiple** option in the mode of **TEXTEDIT** command, AutoCAD will continue to prompt you to select an annotation object. If you have finished editing and do not want to select another attribute object to edit, press ENTER to exit the command.

INSERTING BLOCKS WITH ATTRIBUTES

The value of the attributes can be specified during block insertion, either at the command prompt or in the **Edit Attributes** dialog box, if the system variable **ATTDIA** is set to **1**. When you use the **Insert** tool or the **-INSERT** command to insert a block in a drawing, and after you have specified the insertion point, scale factors, and rotation angle, the **Edit Attributes** dialog box will be displayed.

If the value of the system variable **ATTDIA** is set to **0,** the **Edit Attributes** dialog box will be disabled and the prompts and their default values, which you had specified with the attribute definition, are then displayed at the command prompt after you have specified the insertion point, scale, and rotation angle for the block to be inserted.

In the **Edit Attributes** dialog box, the prompts that were entered at the time of attribute definition in the dialog box are displayed with their default values in the corresponding fields. If an attribute has been defined with the **Constant** mode, it is not displayed in the dialog box because a constant attribute value cannot be edited. You can enter the attribute values in the fields located next to the attribute prompt. If no new values are specified, the default values are displayed. If there are more attributes, they can be accessed by using the **Next** or **Previous** button. The block name is displayed at the top of the dialog box. After entering the new attribute values, choose the **OK** button. AutoCAD will place these attribute values at the specified location.

Tutorial 2

In this tutorial, you will insert the block (COMP) that was defined in Tutorial 1. The following is the list of the attribute values for computers:

ITEM	MAKE	PROCESSOR	HD	RAM
Computer	Gateway	486-60	150 MB	16 MB
Computer	Zenith	486-30	100 MB	32 MB
Computer	IBM	386-30	80 MB	8 MB
Computer	Dell	586-60	450 MB	64 MB
Computer	CAD-CIM	Pentium-90	100 Min	32 MB
Computer	CAD-CIM	Unknown	600 MB	Standard

1. Draw the floor plan shown in Figure 17-4 (assume the dimensions).

Figure 17-4 *Floor plan drawing for Tutorial 2*

2. Choose the **Insert** tool available in the **Blocks** panel in the **Insert** tab to insert the blocks. When you invoke the **Insert** tool, the **Insert** dialog box is displayed. Enter **COMP** in the **Name** edit box and choose the **OK** button to exit the dialog box. Select an insertion point on the screen to insert the block. After you specify the insertion point, the **Edit Attributes** dialog box is displayed. In this dialog box, you can specify the attribute values, if you need to, in the different edit boxes.

3. Choose the **Insert** tool again to insert other blocks and define their attribute values, as shown in Figure 17-5.

4. Save the drawing for further use.

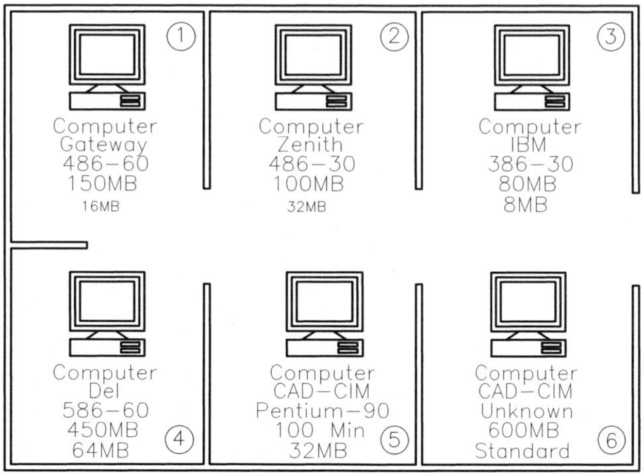

Figure 17-5 *The floor plan after inserting blocks and defining their attributes*

MANAGING ATTRIBUTES

Ribbon: Insert > Block Definition > Manage Attributes **Command:** BATTMAN
Menu Bar: Modify > Object > Attribute > Block Attribute Manager

The **Manage Attributes** tool allows you to manage attribute definitions for blocks in the current drawing. Choose this tool; the **Block Attribute Manager** dialog box will be displayed. If the attribute is edited in the **Constant** mode, then changes in the default value of the attribute in the existing drawing can be seen. If the mode is other than the **Constant** mode, the changes in the attribute can only be seen for new block insertions.

EXTRACTING ATTRIBUTES

Ribbon: Insert > Linking & Extraction > Data Extraction
Command: EATTEXT or DATAEXTRACTION

The **Extract Data** tool allows you to extract the block attribute information and property information such as drawing summary from single or multiple drawings. When this tool is invoked, the **Data Extraction** wizard is displayed.

The ATTEXT Command for Attribute Extraction

Command: ATTEXT

The **ATTEXT** command allows you to use the **Attribute Extraction** dialog box for extracting attributes. The information about the **File Format**, **Template File**, and **Output File** must be entered in the dialog box to extract the defined attribute. Also, you must select the blocks whose attribute values you want to extract. If you do not specify a particular block, all the blocks in the drawing are used.

EXERCISES

Exercise 1 — Attribute Definition

In this exercise, you will define the following attributes for a resistor and then create a block using the **Create Block** tool. The name of the block is **RESIS**. The distance between the dotted lines is 0.5 units.

Mode	Tag name	Prompt	Default value
Verify	RNAME	Enter name	RX
Verify	RVALUE	Enter resistance	XX
Verify, Invisible	RPRICE	Enter price	00

1. Draw the resistor, as shown in Figure 17-6.

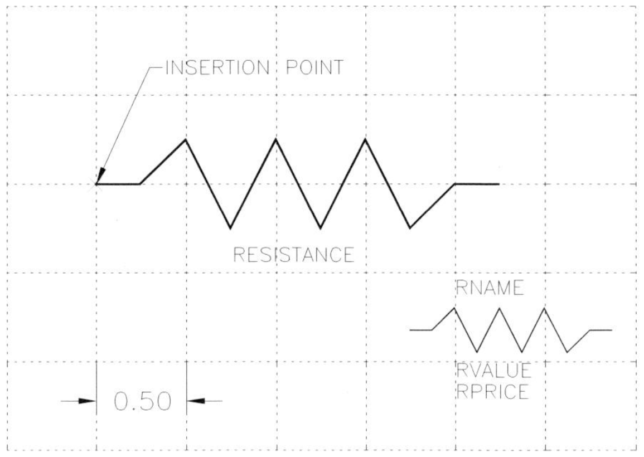

Figure 17-6 *Drawing of a resistor for Exercise 1*

2. Choose the **Define Attributes** tool to invoke the **Attribute Definition** dialog box.

3. Define the attributes, as shown in the preceding table, and position the attribute text as shown in Figure 17-6.

4. Use the **BLOCK** command to create a block. The name of the block is **RESIS**, and the insertion point of the block is at the left end of the resistor. When you select the objects for the block, make sure you also select the attributes.

Exercise 2

In this exercise, you will use the **Insert** tool to insert the block that was defined in Exercise 1 (RESIS). The following is the list of the attribute values for the resistances in the electric circuit:

RNAME	RVALUE	RPRICE
R1	35	.32
R2	27	.25
R3	52	.40
R4	8	.21
RX	10	.21

1. Draw the electric circuit diagram, as shown in Figure 17-7 (assume the dimensions).

2. Set the system variable **ATTDIA** to **1**. Use the **Insert** tool to insert the blocks and define the attribute values in the **Attribute Definition** dialog box.

3. Repeat the **Insert** tool to insert other blocks, and define their attribute values as given in the table. Save the drawing as *attexr2.dwg*, refer to Figure 17-8.

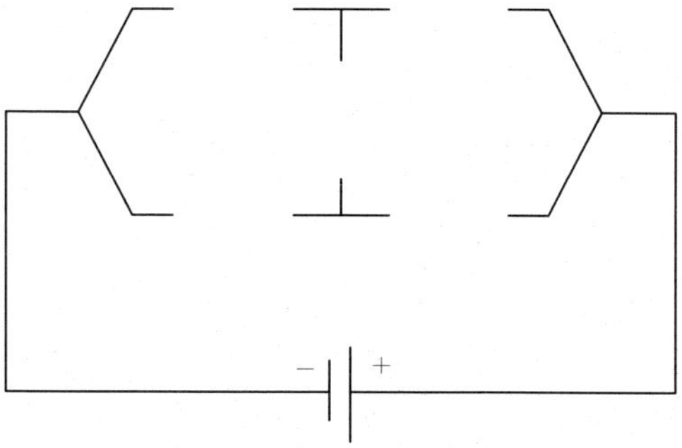

Figure 17-7 *Electric circuit diagram without resistors for Exercise 2*

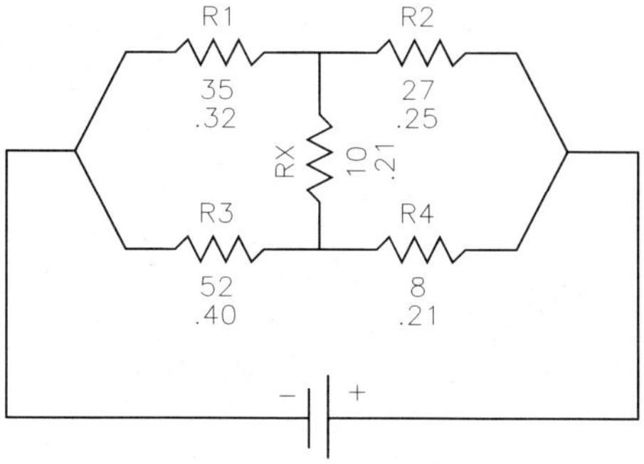

Figure 17-8 *Electric circuit diagram with resistors for Exercise 2*

Chapter 18

Understanding External References

Learning Objectives

After completing this chapter, you will be able to:

- *Understand external references and their applications*
- *Use the External References Palette*
- *Understand the difference between the Overlay and Attachment options*
- *Use of the Attach tool*
- *Work with Underlays*
- *Use the DesignCenter to attach a drawing as an xref*
- *Use the Bind tool to add dependent symbols*
- *Use the Clip tool to clip xref drawings*

EXTERNAL REFERENCES

The external reference feature allows you to reference an external drawing without making that drawing a permanent part of the existing drawing. For example, assume that you have an assembly drawing ASSEM1 that consists of two parts, SHAFT and BEARING. The SHAFT and BEARING are separate drawings created by two CAD operators or provided by two different vendors. You want to create an assembly drawing from these two parts. One way to create an assembly drawing is to insert these two drawings as blocks by using the **Insert** tool in the **Block** panel. Now assume that the design of BEARING has changed due to customer or product requirements. To update the assembly drawing, you have to make sure that you insert the BEARING drawing after the changes have been made. If you forget to update the assembly drawing, then the assembly drawing will not reflect the changes made in the piece part drawing. In a production environment, this could have serious consequences.

You can solve this problem by using the external reference facility, which lets you link the piece part drawings with the assembly drawing. If the xref drawings (piece part) get updated, the changes are automatically reflected in the assembly drawing. This way the assembly drawing stays updated, no matter when the changes were made in the piece part drawings. There is no limit to the number of drawings that you can reference. You can also have nested references. For example, the piece part drawing BEARING could be referenced in the SHAFT drawing, and the SHAFT drawing could be referenced in the assembly drawing ASSEM1. When you open or plot the assembly drawing, AutoCAD automatically loads the referenced drawing SHAFT and the nested drawing BEARING. While using external references, several people working on the same project can reference the same drawing and all the changes made are displayed everywhere the particular drawing is being used.

MANAGING EXTERNAL REFERENCES IN A DRAWING

Ribbon: View > Palettes > External References Palette
Toolbar: Reference > External References **Command:** XREF
Menu Bar: Insert > External References

When you choose the **External References Palette** button from the **Palettes** panel of the **View** tab, refer to Figure 18-1, AutoCAD displays the **EXTERNAL REFERENCES** palette. The **EXTERNAL REFERENCES** palette displays the status of each Xref in the current drawing and the relation between the various Xrefs. It allows you to attach a new xref and detach, unload, and load an existing one, change an attachment to an overlay, or an overlay to an attachment. You can also open a reference drawing for editing from this palette. Additionally, it allows you to edit an xref's path and bind the xref definition to the drawing.

*Figure 18-1 The **Palettes** panel*

THE OVERLAY OPTION

As discussed earlier, when you attach an xref to a drawing, the **Attach External Reference** dialog box is displayed. The **Reference Type** area of this dialog box has two radio buttons, **Attachment** and **Overlay**. You can select any of these radio buttons to xref a drawing. The **Attachment** radio button is selected by default. The advantage of selecting the **Overlay** radio button is that you can access the desired drawing instead of the drawing along with its xreffed attachments.

Tutorial 1 — Attachment and Overlay

In this tutorial, you will use the **Attachment** and **Overlay** options to attach and reference the drawings. Two drawings, PLAN and PLANFORG are given. The PLAN drawing, refer to Figure 18-2, consists of the floor plan layout, and the PLANFORG drawing, refer to Figure 18-3 has the details of the forging section only. The CAD operator who is working on the PLANFORG drawing wants to xref the PLAN drawing for reference. Also, the CAD operator working on the PLAN drawing should be able to xref the PLANFORG drawing to complete the project. The following steps illustrate how to accomplish the defined task without creating a circular reference.

Download these files from *https://www.cadsofttech.com*. The path of the file is as follows: Textbooks>CAD/CAM>AutoCAD>AutoCAD 2019: A Problem-Solving Approach, Basic and Intermediate >Input Files.

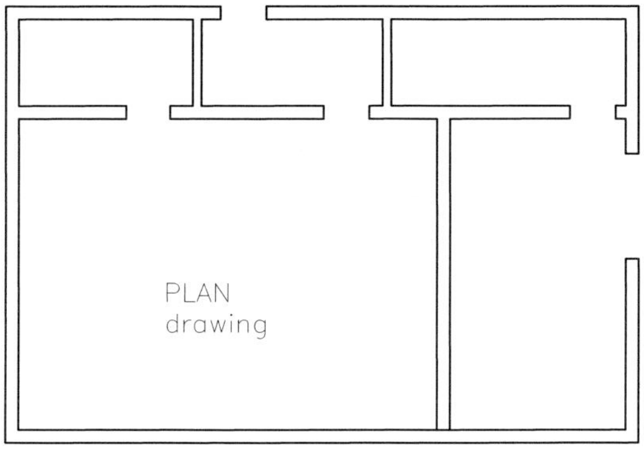

Figure 18-2 *The PLAN drawing*

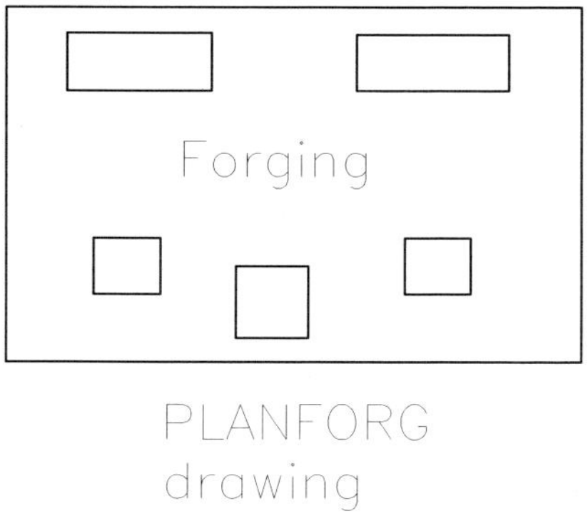

Figure 18-3 *The PLANFORG drawing*

How circular reference is caused?

1. Open the drawing PLANFORG and then choose the **External References Palette** button from the **Palettes** panel in the **View** tab. Next, choose the **Attach DWG** option from the **EXTERNAL REFERENCES** palette; the **Select Reference File** dialog box is displayed. Select the PLAN drawing from the list box of the **Select Reference File** dialog box and choose the **Open** button; the **Attach External Reference** dialog box is displayed. In this dialog box, the name of the PLAN drawing is displayed in the **Name** edit box, and the **Attachment** radio button is selected by default in the **Reference Type** area. Choose the **OK** button to exit the dialog box and specify an insertion point on the screen. Now, the drawing consists of the PLANFORG and PLAN. Save the drawing.

2. Open the drawing file PLAN. Next, choose the **External References Palette** button and attach the PLANFORG drawing to the PLAN drawing using the same steps as described

in Step 1. AutoCAD will display the message that the circular reference has been detected and will ask you if you want to continue. If you choose **Yes** in the AutoCAD message box, the circular reference is broken and you are allowed to reference the specific drawing.

The possible solution for the operator working on the PLANFORG drawing is to detach the PLAN drawing. This way the PLANFORG drawing does not contain any reference to the PLAN drawing and would not cause any circular reference. The other solution is to use the **Overlay** option, as follows :

How to prevent circular reference?

3. Open the PLANFORG drawing, refer to Figure 18-4 and select the **Overlay** radio button in the **Attach External Reference** dialog box, which is displayed after you have selected the PLAN drawing to reference. The PLAN drawing is overlaid on the PLANFORG drawing, refer to Figure 18-5. Save the drawing.

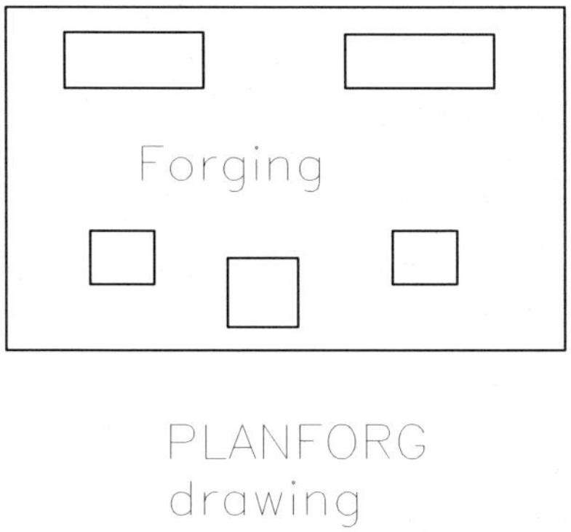

Figure 18-4 *The PLANFORG drawing*

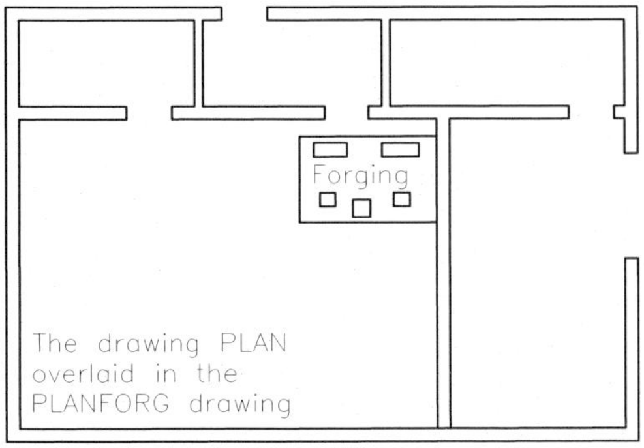

Figure 18-5 *The PLANFORG drawing after overlaying the PLAN drawing*

4. Open the drawing file PLAN, refer to Figure 18-6, and select the **Attachment** radio button in the **Attach External Reference** dialog box, which is displayed when you select the PLANFORG drawing in the **Select Reference File** dialog box to attach it as an xref to the PLAN drawing. You will notice that only the PLANFORG drawing is attached, refer to Figure 18-7. The drawing that was overlaid in the PLANFORG drawing (PLAN) does not appear in the current drawing. This way, the CAD operator working on the PLANFORG drawing can overlay the PLAN drawing, and the CAD operator working on the PLAN drawing can attach the PLANFORG drawing, without causing a circular reference.

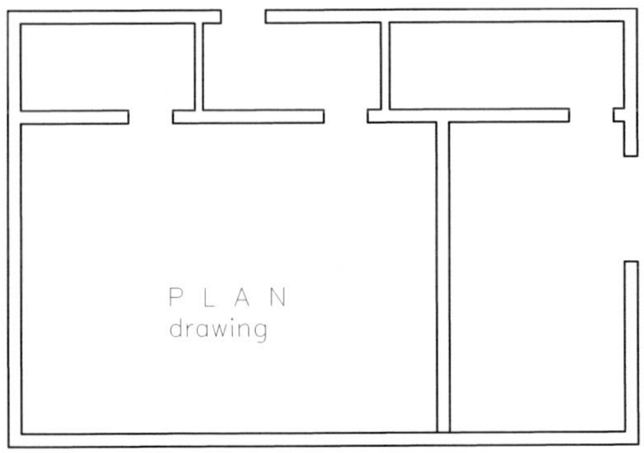

Figure 18-6 The PLAN drawing

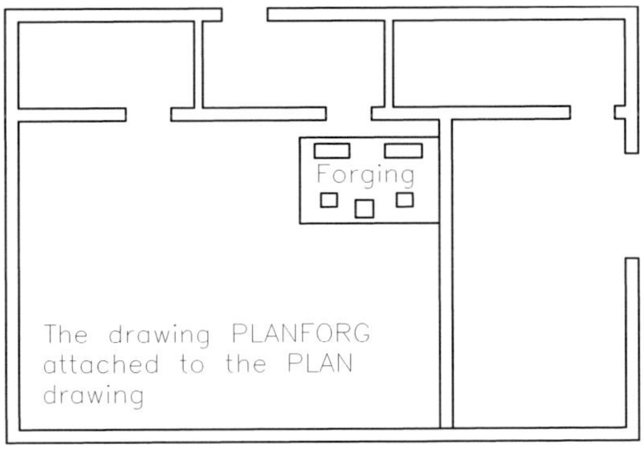

Figure 18-7 The PLAN drawing after attaching the PLANFORG drawing

ATTACHING FILES TO A DRAWING

Ribbon: Insert > Reference > Attach **Command:** ATTACH

You can use the **Attach** tool in the **Reference** panel, refer to Figure 18-8, to attach a DWG, DGN, DWF, PDF, Autodesk Point Cloud files, Navisworks files, or image file without invoking the **EXTERNAL REFERENCES** palette. Using this tool, you can attach a drawing file easily, since most of the xref operations involve simply attaching a drawing file. When you invoke this tool, AutoCAD displays the **Select Reference File** dialog box. To attach a *.dwg, .dgn, .dwf, .pdf, .rcp, rcs, .nwc, .nwd*, or image file, specify it in the **Files of type** drop-down list in the **Select Reference File** dialog box; the corresponding files will be listed in the dialog box. Select the drawing file to be attached and choose the **Open** button; the **Attach External Reference** dialog box is displayed. Select the **Attachment** radio button (if not selected by default) under the **Reference Type** area. You can specify the insertion point, scale, and rotation angle on screen or in the respective edit boxes.

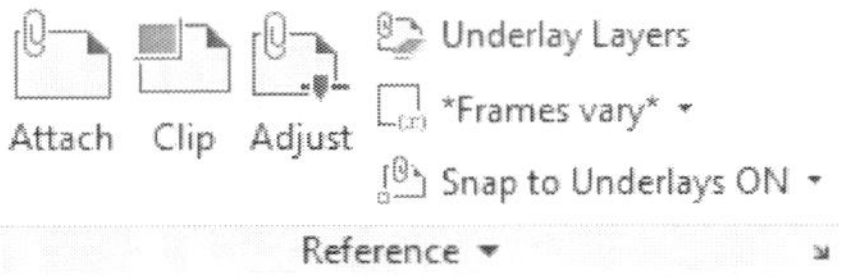

***Figure 18-8** The **Attach** tool in the **Reference** panel*

WORKING WITH UNDERLAYS

You can attach a DWF, DGN, or PDF file as an underlay to the current drawing file. The underlay files are not a part of original drawing files. Therefore, if you add a file as an underlay, it does not increase the file size of the current drawing. The procedure to add a file as an underlay is similar to attaching a drawing file using the **Attach** tool. After you select the file to attach, the **Attach <XXXX> Underlay** dialog box will be displayed, where <XXXX> is the file type. Figure 18-9 shows the **Attach PDF Underlay** dialog box that is displayed on selecting a pdf file from the **Select Reference File** dialog box.

If the selected *.pdf* file has multiple pages, all pages of the pdf file are listed in the **Select one or more pages from the PDF file** area. Select the pages to be attached from this area. If you have selected multiple pages as well as the **Specify on-screen** check box from the **Insertion point** area, then you are prompted to specify different insertion points for different pages.

The files attached as an underlay behave like blocks. The general modify commands like move, copy, rotate, mirror, and so on can be applied on them. However, you cannot bind a file that is attached as an underlay or modify the attached file in the current drawing file.

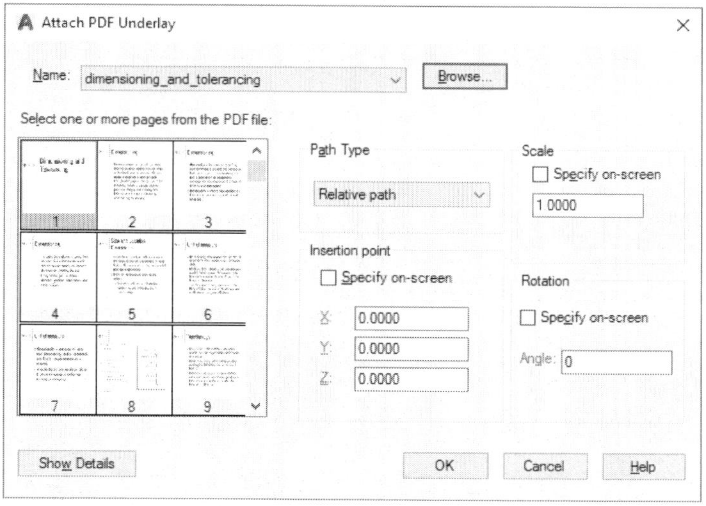

Figure 18-9 *The **Attach PDF Underlay** dialog box*

USING THE DesignCenter TO ATTACH A DRAWING AS AN XREF

The **DESIGNCENTER** can also be used to attach an xref to a drawing. To do so, choose the **DesignCenter** button from the **Palettes** panel in the **View** tab; the **DESIGNCENTER** palette will be displayed. In the **DESIGNCENTER** palette, choose the **Tree View Toggle** button, if it is not chosen already, to display the tree pane. Expand the **Tree view** and double-click on the folder whose contents you want to view. The contents of the selected folder are displayed in the palette. From the thumbnails of drawings displayed on the right-side of the palette, right-click on the drawing to be attached as an xref; a shortcut menu is displayed. Choose **Attach as Xref** from the shortcut menu; the **Attach External Reference** dialog box will be displayed. Alternatively, you can also use the right mouse button to drag and drop the drawing into the current drawing; a shortcut menu will be displayed. Choose the **Create Xref** option to insert the drawing as an xref.

ADDING XREF DEPENDENT NAMED OBJECTS

Menu Bar: Modify > Object > External Reference > Bind
Toolbar: Reference > Xbind **Command:** XB/XBIND

You can use the **Xbind** tool to add the selected named objects such as blocks, dimension styles, layers, line types, and text styles of the xref drawing to the current drawing.

CLIPPING EXTERNAL REFERENCES

Ribbon: Insert > Reference > Clip **Command:** CLIP

The **Clip** tool is used to trim an xref after it has been attached to a drawing to display only a portion of the drawing. After you have attached an xref to a drawing, you can trim it to display only a portion of the drawing by using the **Clip** tool. However, clipping xref does not modify the referenced drawing. On invoking the **Clip** tool, you are prompted to select an object to clip. Select a DWG, DGN, IMAGE, or PDF; the respective clipping options will be displayed.

EXERCISE

Exercise 1 *Bind Xref*

In this exercise, you will start a new drawing and xref the drawings as Part-1 and Part-2, refer to Figure 18-10 and Figure 18-11. For assembly, refer to Figure 18-12. You will also edit one of the piece parts to correct the size and use the **Bind Xref** option to bind some of the dependent symbols to the current drawing. The parameters of layers for Part-1, Part-2 and ASSEM1 are as follows:

For Part-1, set up the following layers:

Layer Name	Color	Linetype
0	White	Continuous
Object	Red	Continuous
Hidden	Blue	Hidden2
Center	White	Center2
Dim-Part1	Green	Continuous

For Part-2, set up the following layers:

Layer Name	Color	Linetype
0	White	Continuous
Object	Red	Continuous
Hidden	Blue	Hidden
Center	White	Center
Dim-Part2	Green	Continuous
Hatch	Magenta	Continuous

For ASSEM1, set up the following layers:

Layer Name	Color	Linetype
0	White	Continuous
Object	Blue	Continuous
Hidden	Yellow	Hidden

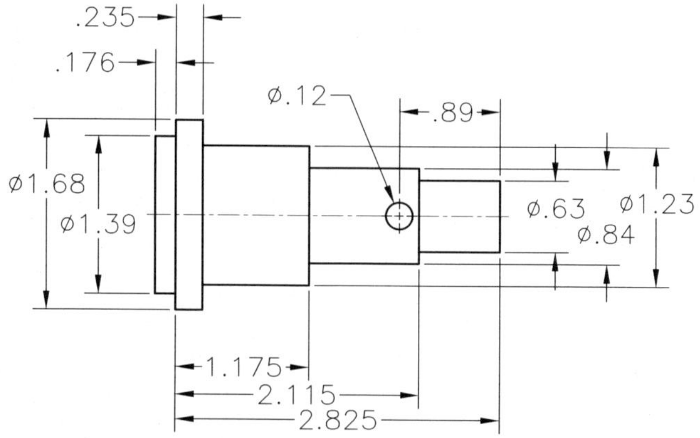

Figure 18-10 *Drawing of Part-1*

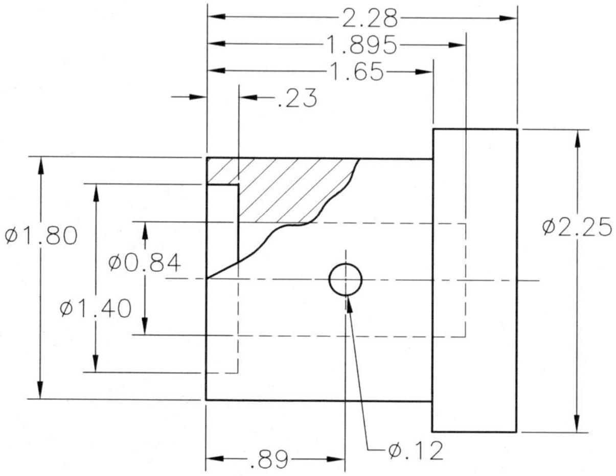

Figure 18-11 *Drawing of Part-2*

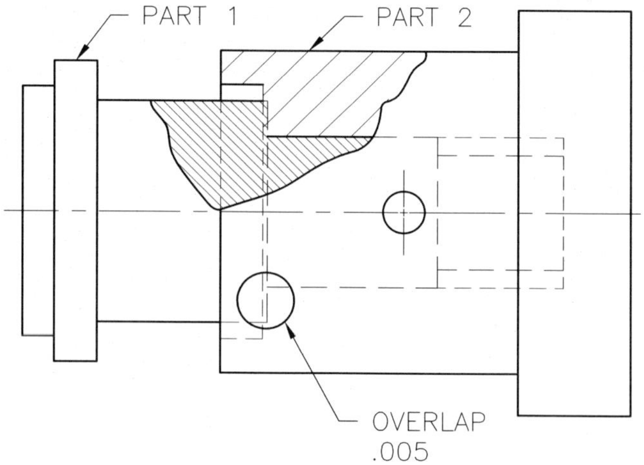

Figure 18-12 *Assembly drawing after attaching Part-1 and Part-2*

Chapter 19

Working with Advanced Drawing Options

Learning Objectives

After completing this chapter, you will be able to:

- *Define multiline style and specify the properties of multilines using the Multiline Style command*
- *Draw various types of multilines using the MLINE command*
- *Edit multilines using the MLEDIT command*
- *Draw NURBS splines using the Spline tool*
- *Edit NURBS splines using the Edit Spline tool*
- *Compare Drawings*

UNDERSTANDING THE USE OF MULTILINES

The Multiline feature allows you to draw composite lines that consist of multiple parallel lines. These parallel lines are called elements. You can draw these lines using the **MLINE** command. But before drawing multilines, you need to set the multiline styles. This can be accomplished by using the **MLSTYLE** command. You can also edit multilines by using the **MLEDIT** command.

DEFINING A MULTILINE STYLE

Menu Bar: Format > Multiline Style **Command:** MLSTYLE

You can create a multiline style by using the **MLSTYLE** command. You can specify the number of elements in a multiline and the properties of each element. The style also controls the start and end caps, the start and end lines, and the color of multilines and fill. When you invoke the **Multiline Style** tool from the **Format** menu or by using the **MLSTYLE** command, AutoCAD displays the **Multiline Style** dialog box, as shown in Figure 19-1. Using this dialog box, you can set the spacing between parallel lines, specify linetype pattern, set colors, solid fill, and capping arrangements. By default, the multiline style (STANDARD) has two lines that are offset at 0.5 and -0.5.

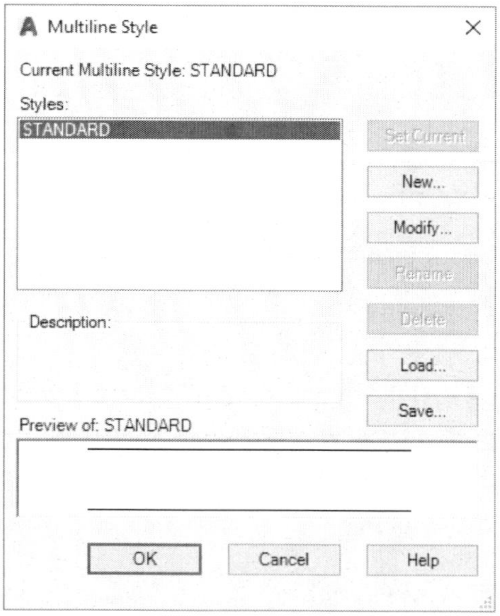

Figure 19-1 *The* ***Multiline Style*** *dialog box*

DRAWING MULTILINES

Menu Bar: Draw > Multiline **Command:** MLINE/ML

The **MLINE** command is used to draw multilines. The following is the prompt sequence for the **MLINE** command:

Specify start point or [Justification/Scale/STyle]: *Select a start point or enter an option.*
Specify next point: *Select the second point.*

Specify next point or [Undo]: *Select the next point.*
Specify next point or [Close/Undo]: *Select the next point, enter **U**, or enter **C** for close.*

When you invoke the **MLINE** command, it always displays the current status of the multiline justification, scale, and style name. Remember that all line segments created in a single **MLINE** command are a single entity.

EDITING MULTILINES BY USING GRIPS

Multilines can be easily edited using grips. When you select a multiline, the grips appear at the endpoints, based on the justification used while drawing multilines. For example, if the multilines are top-justified, the grips will be displayed at the endpoint of the first (top) line segment. Similarly, for zero and bottom-justified multilines, the grips are displayed on the centerline and bottom line, respectively.

EDITING MULTILINES BY USING DIALOG BOX

Menu Bar: Modify > Object > Multiline **Command:** MLEDIT

When you invoke the **MLEDIT** command, AutoCAD displays the **Multilines Edit Tools** dialog box. This dialog box contains five basic editing tools. To edit a multiline, first select the editing operation you want to perform by clicking on the image tile. Once you have selected the editing option, AutoCAD will prompt you to select the first and second multilines. After editing one set of multilines, you will be prompted again to select the first multiline or undo the last editing. Press ENTER to exit this command after you have finished the editing.

Tutorial 1 — *Multiline*

In this tutorial, you will create a multiline style that represents a wood-frame wall system. The wall system consists of ½" wallboard, 4 X ½" wood stud, and ½" wallboard.

1. Enter **MLSTYLE** at the command prompt to display the **Multiline Style** dialog box. Also, the current style **STANDARD** is displayed in the **Styles** list box.

2. Choose the **New** button to invoke the **Create New Multiline Style** dialog box. Enter **MYSTYLE** in the **New Style Name** edit box and choose **Continue** to display the **New Multiline Style: MYSTYLE** dialog box.

3. Enter **Wood-frame Wall System** in the **Description** edit box.

4. Select the **0.5** line definition in the **Elements** list box. In the **Offset** edit box, replace **0.500** with **1.00**. This redefines the first line which will now be 1.00" above the centerline of the wall.

5. Select the **-0.5** line definition in the **Elements** list box. In the **Offset** edit box, replace **-0.500** with **-1.00**. This redefines the second line which will now be placed 1.00" below the centerline of the wall.

6. Choose the **Add** button; a new line is added with offset value of 0 and is highlighted in the **Elements** list box.

7. In the **Offset** edit box, replace **0.000** with **1.50**.

8. Select **Yellow** from the **Color** drop-down list.

9. Repeat steps 6 through 8, but this time use the value **-1.50** in place of **1.50** in step 7 to add another line to the current multiline style. The color of this line should be red.

10. Choose the **OK** button and return to the **Multiline Style** dialog box. The new multiline style is displayed in the **Preview of** area.

11. Choose the **Set Current** button to make the **MYSTYLE** as current style.

12. Choose the **OK** button from the **Multiline Style** dialog box to return to the drawing area. To test the new multiline style, use the **MLINE** command and draw a series of lines, refer to Figure 19-2.

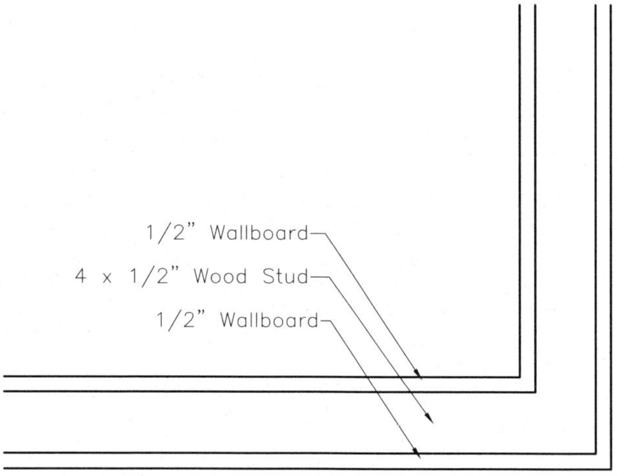

Figure 19-2 *Multiline style created for Tutorial 1*

CREATING REVISION CLOUDS

Ribbon: Home > Draw > Revision Cloud drop-down or Annotate > Markup > Revision Cloud drop-down
Tool Palettes: Draw > Revision Cloud
Toolbar: Draw > Revision Cloud
Menu Bar: Draw > Revision Cloud
Command: REVCLOUD

Revision cloud is the sequential arc of polylines and is used to create a cloud shaped object. Revision clouds can be used to highlight the details of a drawing. The tools to draw a revision cloud are grouped in the **Revision Cloud** drop-down of the **Draw** panel in the **Ribbon**. Remember that the tool that was used last to create a revision cloud will be displayed in this drop-down. The different methods to draw a revision cloud are discussed next.

Rectangular Revision Cloud

To draw a rectangular revision cloud, choose the **Rectangular Revision Cloud** tool from the **Revision Cloud** drop-down in the **Draw** panel; you will be prompted to specify the first point. Specify the first point or enter coordinates. Next, specify the opposite corner point of the rectangle.

Polygonal Revision Cloud

To draw a polygonal revision cloud, choose the **Polygonal Revision Cloud** tool from the **Draw** panel. You will be prompted to specify the first point, specify the first point; you will be prompted to specify the next point. Keep specifying the points until the desired polygonal revision cloud is created. Press ENTER to terminate the command.

Freehand Revision Cloud

To draw a freehand revision cloud, choose the **Freehand Revision Cloud** tool from the **Draw** panel. You will be prompted to specify the first point; specify the first point. Now, as you move the cursor, different arcs of the cloud with varied radii are drawn. When the start point and endpoint meet, the revision cloud is completed and you get a message that the revision cloud has been created.

CREATING WIPEOUTS

Ribbon: Home > Draw > Wipeout or Annotate > Markup > Wipeout
Command: WIPEOUT

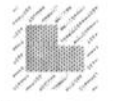

The **Wipeout** tool is used to create a polygonal area to cover the existing objects with the current background color. The area defined by this tool is governed by the wipeout frame. The frame can be turned on and off for editing and plotting the drawings, respectively. This tool can be used to add notes and details to a drawing.

CREATING NURBS

Ribbon: Home > Draw > Spline Fit **Menu Bar:** Draw > Spline > Fit Points
Tool Palettes: Draw > Spline **Toolbar:** Draw > Spline **Command:** SPLINE/SPL

The NURBS is an acronym for **Non-Uniform Rational Bezier-Spline**. These splines are considered true splines. In AutoCAD, you can create NURBS using the **Spline** tool. The spline created with the **Spline** tool is different from the spline created using the **Polyline** tool. The non-uniform aspect of the spline enables the spline to have sharp corners because the spacing between the spline elements that constitute a spline can be irregular. Rational means that irregular geometry such as arcs, circles, and ellipses can be combined with free-form curves. The Bezier-spline (B-spline) is the core that enables accurate fitting of curves to input data with Bezier's curve-fitting interface. Not only are spline curves more accurate compared to smooth polyline curves, but they also use less disk space.

EDITING SPLINES

Ribbon: Home > Modify > Edit Spline **Menu Bar:** Modify > Object > Spline
Tool Palettes: Modify > Edit spline **Toolbar:** Modify II > Edit Spline
Command: SPLINEDIT /SPE

The NURBS can be edited using the **Edit Spline** tool. Using this tool, you can fit data in a selected spline, close or open a spline, move vertex points, and refine or reverse a spline. Apart from the ways mentioned in the preceding command box, you can choose the options available in the **Spline** cascading menu that will be displayed when you select a spline and right-click. The prompt sequence that will follow when you choose the **Edit Spline** tool in the **Modify** panel is given next.

Select spline: *Select the spline to be edited if not selected already using the above-mentioned shortcut menu.*
Enter an option [Close/Join/Fit data/Edit vertex/convert to Polyline/Reverse/Undo/eXit] <eXit>: *Select any one of the options.*

EDITING SPLINES USING 3D EDIT BAR

Command: 3DEDITBAR

The **3DEDITBAR** command is used to edit a spline by using the **Move** gizmo. To edit a spline using this command, select the spline and right click; a shortcut menu will be displayed. Choose the **Spline > 3D Edit Bar** option from the shortcut menu; you will be prompted to select a point on the curve. Select a point on the curve at the position where you want to modify the spline; the Move gizmo will be displayed with one extra arrow tangential to the curve at the selected point. This arrow can be used to change the tangent direction as well as the magnitude of tangency at the point. Using this arrow, you can dynamically edit the spline.

DWG Compare

New

Ribbon: Collaborate > Compare > DWG Compare **Command:** COMPARE

The **DWG Compare** tool is used to compare two drawings with each other. It highlights the differences between two revisions of the same drawing or different drawings. This tool can be activated with or without any drawings open. If no drawings are open, the **DWG Compare** tool can be accessed from the Application Menu.

EXERCISES

Exercise 1

Create the drawing shown in Figure 19-3.

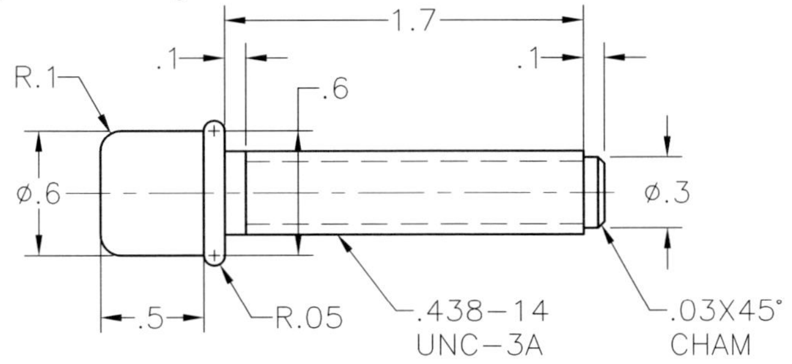

Figure 19-3 *Drawing for Exercise 1*

Exercise 2

Create the drawing shown in Figure 19-4. Use the **MLINE** command to draw the walls. The wall thickness is 12 inches. Assume the missing dimensions.

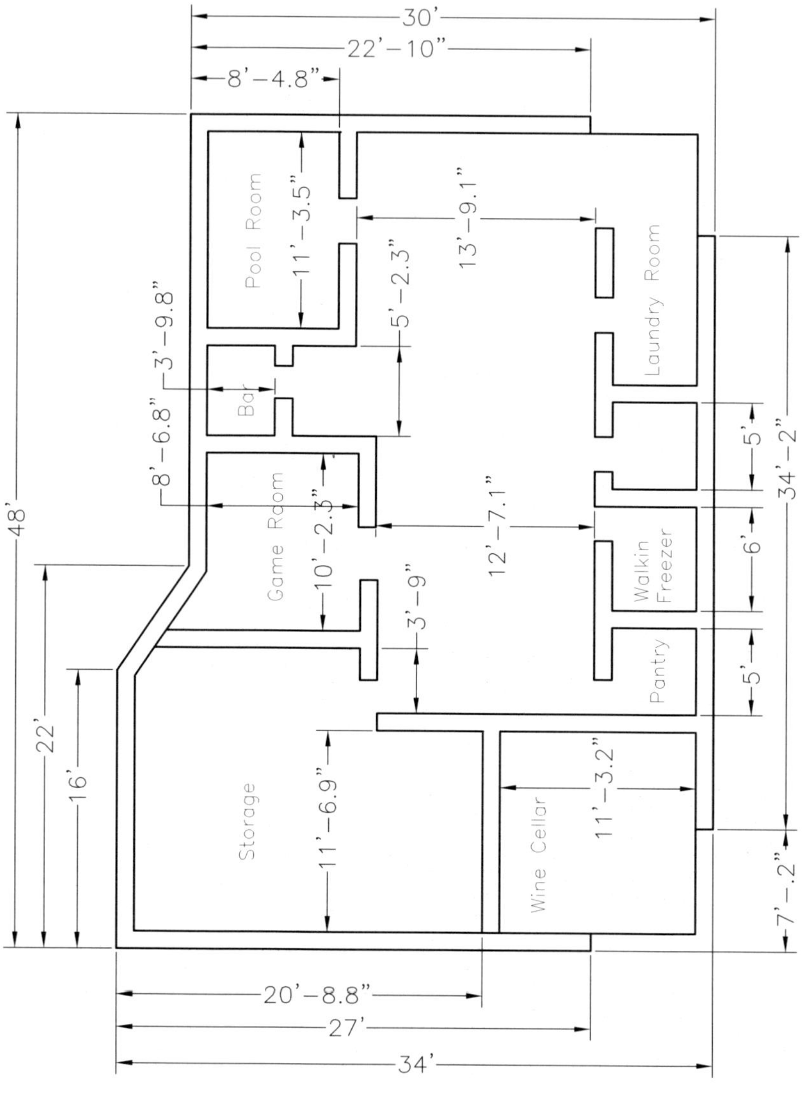

Figure 19-4 *Drawing for Exercise 2*

Chapter 20

Grouping and Advanced Editing of Sketched Objects

Learning Objectives

After completing this chapter, you will be able to:

- *Group sketched objects*
- *Select and cycle through defined groups*
- *Change properties and location of sketched objects*
- *Perform editing operations on polylines*
- *Explode compound objects and undo previous commands*
- *Rename the named objects and remove unused named objects*

GROUPING SKETCHED OBJECTS USING THE OBJECT GROUPING DIALOG BOX

Ribbon: Home > Groups > Group Manager **Command:** CLASSICGROUP

In AutoCAD, you can group sketched objects using the **Object Grouping** dialog box, as shown in Figure 20-1. To invoke the **Object Grouping** dialog box, you need to choose the **Group Manager** tool from the **Groups** panel in the **Home** tab. You can use the **Object Grouping** dialog box to group AutoCAD objects and assign a name to the group. Once you have created groups, you can select the objects by the group name. The individual characteristics of an object are not affected by forming groups. Groups enable you to select all the objects in a group together for editing. It makes the object selection process easier and faster. Objects can be members of several groups. Also, a group can contain several smaller groups. This may be referred to as nested groups.

Although an object belongs to a group, you can still select an object as if it does not belong to any group. Groups can be selected by entering the group name or by selecting an object that belongs to the group. You can also highlight the objects in a group or sequentially highlight the groups to which an object belongs. The options in the **Object Grouping** dialog box are discussed next.

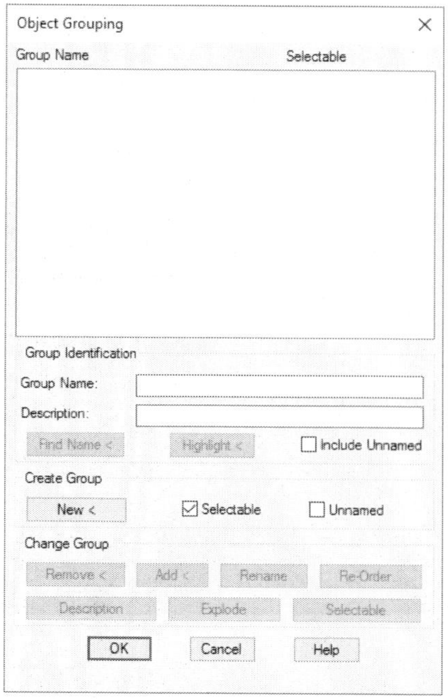

Figure 20-1 *The* ***Object Grouping*** *dialog box*

Tutorial 1 Reorder Group

In this tutorial, you will draw a rectangle representing a toolpath. The rectangle should comprise of four individual lines. Group all lines by selecting them in the order shown in Figure 20-2(a). Then, highlight the order, and re-order the group, as shown in Figure 20-2(b).

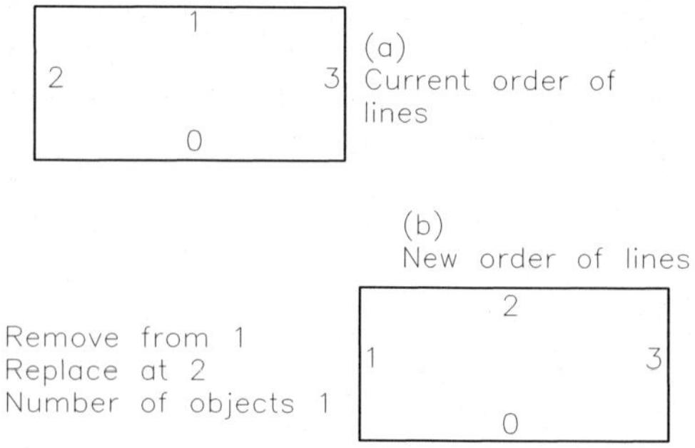

Figure 20-2 *Changing the order of group objects*

1. Draw four lines that are connected at the end points to form a rectangle.

2 Choose the **Group Manager** tool from the **Groups** panel; the **Object Grouping** dialog box is displayed.

3. Type **G1** in the **Group Name** edit box and choose the **New** button; the **Object Grouping** dialog box will disappear and you will be prompted to select the objects for grouping.

4. Select the lines in the order specified in Figure 20-2(a) and right-click; the **Object Grouping** dialog box is displayed again.

5. Make the group (G1) current by selecting it in the **Group Name** list box and choose the **Re-Order** button in the dialog box; the **Order Group** dialog box is displayed.

6. Select the group G1 in the **Group Name** list box, if not already selected, and choose the **Highlight** button; the **Object Grouping** dialog box is displayed, as shown in Figure 20-3. You can use the **Next** and **Previous** buttons to highlight the grouped objects in the order of selection.

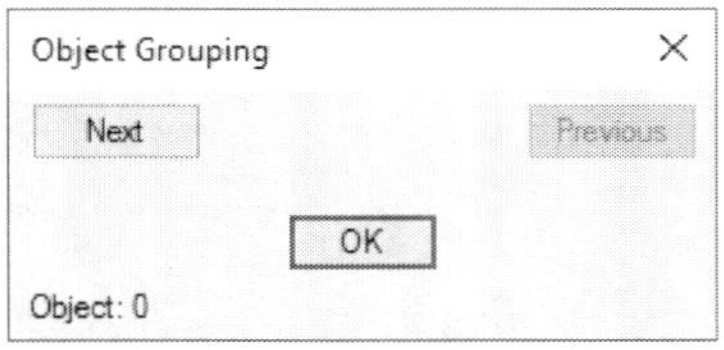

Figure 20-3 *The **Object Grouping** dialog box*

You will notice that the order of selection is not proper. As these lines represent a tool path, you need to reorder the objects, as shown in Figure 20-2(b).

7. To get a clockwise tool path, you must switch object numbers 1 and 2. You can do so by entering the necessary information in the **Order Group** dialog box. Enter **1** in the **Remove from position (0-3)** edit box and **2** in the **Enter new position number for the object (0-3)** edit box. Enter **1** in the **Number of objects (1-4)** edit box because there is only one object to be replaced.

8. After entering the information, choose the **Re-Order** button to define the new order of the objects. You can confirm the change by choosing the **Highlight** button again and cycling through the objects.

GROUPING SKETCHED OBJECTS USING THE GROUP BUTTON

Ribbon: Home > Groups > Group **Command:** GROUP/G
Toolbar: Group > Group

In AutoCAD, you can group objects by using the **Group** tool from the **Groups** panel of the **Home** tab. The prompt sequence to create a group is given next:

*Choose the **Group** tool.*
Select objects or [Name/Description]: **N** Enter
Enter a group name or [?]: *Enter the group name* Enter
Select objects or [Name/Description]: *Select the objects to be grouped.*
Select objects or [Name/Description]: Enter
Group "*group name*" has been created.

SELECTING GROUPS

You can select a group by name by entering **G** at the **Select objects** prompt. For example, if you have to move a particular group, choose the **Move** tool from the **Modify** panel in the **Home** tab and then enter the following prompt sequence:

Select objects: **G** Enter
Enter group name: *Enter the group name*
n found
Select objects: Enter

CHANGING PROPERTIES OF AN OBJECT

AutoCAD provides you different options for changing the properties of an object. These options are discussed next.

Using the PROPERTIES Palette

Ribbon: View > Palettes > Properties **Toolbar:** Standard > Properties
Quick Access Toolbar: Properties (*Customize to Add*) **Command:** PROPERTIES/PR/CH

The categories displayed in the **PROPERTIES** palette depend on the type of object selected. The **General** category displays the general properties of objects, such as color, layer, linetype, linetype scale, plot style, lineweight, hyperlink, and thickness. Depending on the type of object selected, the **Geometry** category will contain a set of different properties. If you have selected many types of objects in a drawing, the drop-down list at the top of the **PROPERTIES** palette will display all the type of selected objects. If you select a type of object from the drop-down list, the corresponding categories of the object properties will be displayed in the palette. You can also invoke the **PROPERTIES** palette from the shortcut menu displayed on selecting the object.

EXPLODING COMPOUND OBJECTS

Ribbon: Home > Modify > Explode **Toolbar:** Modify > Explode
Menu Bar: Modify > Explode **Command:** EXPLODE/X

The **Explode** tool is used to split compound objects such as blocks, polylines, regions, polyface meshes, polygon meshes, multilines, 3D solids, 3D meshes, bodies, or dimensions into the basic objects that make them up. For example, if you explode a polyline or a 3D polyline, the result will be ordinary lines or arcs (tangent specification and width are not considered). When a 3D polygon mesh is exploded, the result is 3D faces. Polyface meshes are turned into 3D faces, points, and lines. Upon exploding 3D solids, the planar surfaces of the 3D solid turn into regions, and non-planar surfaces turn into bodies, multilines are changed to lines. Regions turn into lines, ellipses, splines, or arcs. On exploding, 2D polylines lose their width and tangent specifications and 3D polylines explode into lines. When a body is exploded, it changes into single-surface bodies, curves, or regions. When a leader is exploded, the components are lines, splines, solids, block inserts, text or multiline text, tolerance objects, and so on. Multiline text explodes into a single line text. This tool is especially useful when you have inserted an entire drawing and you need to alter a small detail. After you invoke the **Explode** tool, you are prompted to select the objects you want to explode. After selecting the objects, press ENTER or right-click to explode the selected objects and then end the command.

EDITING POLYLINES

A polyline can assume various characteristics such as width, linetype, joined polyline, and closed polyline. You can edit polylines, polygons, or rectangles to attain the desired characteristics using the **Edit Polyline** tool. In this section, we will be discussing how to edit simple 2D polylines. The following are the editing operations that can be performed on an existing polyline using the **Edit Polyline** tool.

Editing Single Polyline

Ribbon: Home > Modify > Edit Polyline **Toolbar:** Modify II > Edit Polyline
Menu Bar: Modify > Object > Polyline **Command:** PEDIT/PE

You can edit a single polyline by using the **Edit Polyline** tool. To invoke this tool, choose the **Edit Polyline** tool from the expanded **Modify** panel of the **Home** tab; the following prompt sequence is displayed:

Select polyline or [Multiple]:

If the selected entity is not a polyline, the following message will be displayed at the command prompt.

Object selected is not a polyline.
Do you want to turn it into one? <Y>:

If you want to turn the object into a polyline, respond by entering Y or by simply pressing ENTER. To let the object be as it is, enter N. If you enter N, then AutoCAD will prompt you to select another polyline or object to edit. As mentioned earlier, you can avoid this prompt by setting the value of the **PEDITACCEPT** variable to **1**. The subsequent prompts and editing options depend on the type of polyline that has been selected. AutoCAD provides you the option of either selecting a single polyline or multiple polylines. In this case, a single 2D polyline is selected, and its prompt sequence is given next.

Enter an option [Close/Join/Width/Edit vertex/Fit/Spline/Decurve/Ltype gen/Reverse/Undo]: *Enter an option or press ENTER to end command.*

Editing Multiple Polylines

Selecting the **Multiple** option of the **Edit Polyline** tool allows you to select more than one polyline for editing. You can select the polylines using any of the objects selection techniques. After the objects to be edited are selected, press ENTER or right-click to proceed with the command.

UNDOING PREVIOUS COMMANDS

Quick Access Toolbar: Undo **Toolbar:** Standard > Undo **Command:** UNDO/U

In AutoCAD, their are many tools with the **Undo** option, which can be used to undo (nullify) the changes made within these commands. The **Undo** option is used to undo a previous command or to undo more than one command at a time. This command can be invoked by entering **UNDO** at the command prompt. **Undo** is also available in the **Quick Access Toolbar**, the **Edit** menu, and the **Standard** toolbar. The **Undo** tool in the **Quick Access Toolbar** can only undo the previous command and only one command at a time.

REVERSING THE UNDO OPERATION

Quick Access Toolbar: Redo **Toolbar:** Standard > Redo **Command:** REDO

If you right-click in the drawing area, a shortcut menu is displayed. Choose **Redo** to invoke the **REDO** command. The **REDO** command brings back the process you removed previously using the **U** and **UNDO** commands. This command undoes the last **UNDO** command performed, provided it is entered immediately after the **UNDO** command. On using this command, the objects previously undone reappear on the screen.

RENAMING NAMED OBJECTS

Menu Bar: Format > Rename **Command:** RENAME

You can edit the names of the named objects such as blocks, dimension styles, layers, linetypes, styles, UCS, views, and viewports using the **Rename** dialog box. You can select the named object from the list in the **Named Objects** area of the dialog box. The corresponding names of all the objects of the specified type that can be renamed are displayed in the **Items** area.

REMOVING UNUSED NAMED OBJECTS

Application Menu: Drawing Utilities > Purge
Menu Bar: File > Drawing Utilities > Purge **Command:** PURGE

This is another editing operation used for deletion and it was discussed earlier in relation to blocks. You can delete unused named objects such as blocks, layers, dimension styles, linetypes, text styles, shapes, and so on with the help of the **PURGE** tool. When you create a new drawing or open an existing one, AutoCAD records the named objects in that drawing and notes other drawings that reference the named objects. Usually only a few of the named objects in the drawing (such as layers, linetypes, and blocks) are used. For example, when you create a new drawing, the prototype drawing settings may contain various text styles, blocks, and layers which you do not want to use. Also, you may want to delete particular unused named objects such as unused blocks in an existing drawing. Deleting inactive named objects is important and useful because doing so reduces the space occupied by the drawing. With the **PURGE** tool, you can select the named objects you want to delete. You can invoke this tool at any time in the drawing session.

SETTING SELECTION MODES USING THE OPTIONS DIALOG BOX

Application Menu: Options **Command:** OPTIONS/OP

When you select a number of objects, the selected objects form a selection set. You can specify the objects to be selected in the **Options** dialog box which is invoked by choosing the **Options** button from the **Application Menu**. Five selection modes are provided in the **Selection** tab of this dialog box. You can select any one of these modes or a combination of various modes.

EXERCISES

Exercise 1 *Properties*

Draw a hexagon on OBJ layer in red color. Let the linetype be hidden. Now, use the **PROPERTIES** palette to change the layer of the hexagon to some other existing layer, the color to yellow, and the linetype to continuous. Also, in the **PROPERTIES** palette, under the **Geometry** category, specify a vertex in the **Vertex** field or select one using the arrow (Next or Previous) buttons and then relocate it. The values of the **Vertex X** and **Vertex Y** fields change as the coordinate values of the vertices change. Also, notice the change in the value in the **Area** field. You can also assign a start and end lineweight to each of the hexagon segments between the specified vertices. Use the **Tools > Inquiry > List** tool from the menu bar to verify that the changes have taken place.

Exercise 2 *Properties and Edit Polyline*

Draw part (a) in Figure 20-4 and then using the **Properties** and relevant **Edit Polyline** tool options, convert it into parts (b), (c), and (d). The linetype used in part (d) is HIDDEN.

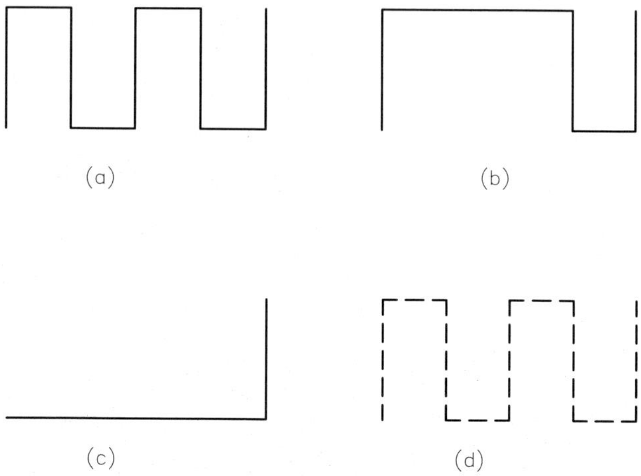

Figure 20-4 *Drawing for Exercise 2*

Chapter 21

Working with Data Exchange & Object Linking and Embedding

Learning Objectives

After completing this chapter, you will be able to:

- *Import and export .dxf files using the Save As and Open tools*
- *Import files*
- *Convert scanned drawings into the drawing editor using the DXB file format*
- *Attach the raster images to the current drawing*
- *Edit raster images*
- *Use DWG Convert and CONTENT EXPLORER*

UNDERSTANDING THE CONCEPT OF DATA EXCHANGE IN AutoCAD

Different companies have developed different softwares for applications such as CAD, desktop publishing, and rendering. This non-standardization of software has led to the development of various data exchange formats that enable the transfer (translation) of data from one data processing software to the other. This chapter will cover various data exchange formats provided in AutoCAD. AutoCAD uses the *.dwg* format to store drawing files. This format is not recognized by some other CAD softwares. To address this problem, AutoCAD provides various data exchange formats such as DXF (Data Interchange Format) and DXB (Drawing Interchange Binary).

CREATING DATA INTERCHANGE (DXF) FILES

The DXF file format generates a text file in ASCII code from the original drawing. This allows any computer system to manipulate (read/write) data in a DXF file. Usually, the DXF format is used for CAD packages based on microcomputers. For example, packages like SmartCAM use DXF files. Some desktop publishing packages, such as PageMaker and Ventura Publisher also use DXF files.

Creating a Data Interchange File

Using the **Save** or **Save As** tool, you can create an ASCII file with a *.dxf* extension from an AutoCAD drawing file. Once you invoke any of these tools, the **Save Drawing As** dialog box will be displayed. Next, choose the **Options** option from the **Tools** drop-down to display the **Saveas Options** dialog box. In this dialog box, the **DWG Options** tab is chosen by default. Select the required *.dxf file extension type from the **Save all drawings as** drop-down list. Next, choose the **DXF Options** tab and enter the degree of accuracy for the numeric values. The default value for the degree of accuracy is sixteen decimal places. You can enter a value between 0 and 16 decimal places.

Importing CAD Files

Ribbon: Insert > Import > Import drop-down > Import
Menu Bar: File > Import **Command:** IMPORT

You can import any surface, solid, 2D and 3D wire geometry from the supported file formats to AutoCAD. To import a CAD geometry to AutoCAD, choose the **Import** tool from the **Import** drop-down of the **Import** panel; the **Import File** dialog box will be displayed. From the **Files of type** drop-down list, select the required format. Next, browse to the required file and then choose the **Open** button from the dialog box; the **Import - Processing Background Job** message box will be displayed stating that the import job is processing in the background. Next, close the message box. The details of the progress can be seen on the bottom-right corner of the screen. After the processing gets completed, a link will be displayed as pop-up. Click on the link to import the CAD geometry into AutoCAD; the inserted part geometry will be placed as a block reference.

OTHER DATA EXCHANGE FORMATS

The other formats that can be used to exchange data from one data processing format to the other are discussed next.

DXB File Format

Menu Bar: Insert > Drawing Exchange Binary **Command:** DXBIN

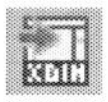
AutoCAD offers another file format for data interchange, DXB. This format is much more compressed than the binary DXF format and is used when you want to translate a large amount of data from one CAD system to another.

Creating and Using an ACIS File

Application Menu: Export > Other Formats **Command:** EXPORT, ACISOUT

Trimmed NURBS, regions, and solids can be exported to an ACIS file with an ASCII format. To do so, choose **Export > Other Formats** from the **Application Menu**; the **Export Data** dialog box will be displayed. Enter a file name in the **File name** edit box. From the **Files of type** drop-down list, select **ACIS (*.sat)** and then choose the **Save** button; you will be prompted to select the objects. Select the solids, regions, or ACIS bodies that you want to add to the file and press ENTER. AutoCAD appends the file extension *.sat* to the file name.

Importing PDF Files

Ribbon: Insert > Import > Import drop-down > PDF Import **Command:** PDFIMPORT

You can import the geometry, raster images, fills, and true type text objects from a PDF file into the current drawing as AutoCAD objects. To do so, choose the **PDF Import** tool from the **Import** drop-down of the **Import** panel in the **Insert** tab; the **Select PDF File** dialog box will be displayed. Select a PDF file to import and choose the **Open** button; the **Import PDF** dialog box will be displayed.

Enter the page number in the **Page** edit box under the **Page to import** area of the **Import PDF** dialog box. To specify the scale, rotation and point of insertion of the imported geometry choose the respective options under the **Location** area. You can also choose the type of data to be imported using the options available in the **PDF data to import** area. To specify the layers in which the PDF data is to be imported, choose the options under the **Layers** area.

Importing 3D Studio Files

Menu Bar: Insert > 3D Studio **Command:** 3DSIN

You can import 3D geometry, views, lights, and objects with surface characteristics created in 3D Studio and saved in *.3ds* format. To do so, choose the **3D Studio** tool from the **Insert** menu; the **3D Studio File Import** dialog box will be displayed. Select the file you wish to import and choose the **Open** button. You can use the **3DSIN** command to import 3D Studio file.

Creating and Using a Windows Metafile

Application Menu: Export > Other Formats **Command:** EXPORT, WMFOUT

The Windows Metafile format (WMF) file contains screen vector and raster graphic formats. In the **Export Data** dialog box or in the **Create WMF File** dialog box, enter the file name. Select **Metafile (*.wmf)** from the **Files of type** drop-down list. The extension *.wmf* is appended to the file name. Next, save the settings and then select the objects you want to save in this file format.

Creating a BMP File

Application Menu: Export > Other Formats **Command:** EXPORT, BMPOUT

This is used to create bitmap images of the objects in your drawing. In the **Export Data** dialog box, enter the name of the file, select **Bitmap (*.bmp)** from the **Files of type** drop-down list and then choose the **Save** button. Select the objects you wish to save as bitmap and press ENTER.

RASTER IMAGES

A raster image consists of small square-shaped dots known as pixels. In a colored image, the color is determined by the color of pixels. The raster images can be moved, copied, or clipped.

Attaching Raster Images

Ribbon: Insert > Reference > Attach **Tool Palettes:** Draw > Attach Image
Toolbar: Reference > Attach Image or Insert > Attach Image **Command:** IMAGEATTACH

Attaching raster images to a drawing does not make them part of the drawing. To attach an image, choose the **Attach** tool from the **Reference** panel; the **Select Reference File** dialog box will be displayed. Select the image to be attached; a preview of the selected image will be displayed in the **Preview** area of the dialog box. Next, choose the **Open** button; the **Attach Image** dialog box will be displayed. The name of the selected image is displayed in the **Name** drop-down list. You can select another file by using the **Browse** button. The **Name** drop-down list displays the names of all the images in the current drawing. Select the **Specify on-screen** check boxes to specify the **Insertion point**, the **Scale**, and **Rotation** angle on the screen. Alternatively, you can clear these check boxes and enter values in the respective edit boxes. On choosing the **Show Details** button, the **Attach Image** dialog box is expanded and it provides the image information such as the Horizontal and Vertical resolution values, current AutoCAD unit, and Image size in pixels and units. Choose the **OK** button to return to the drawing screen.

Managing Raster Images

Ribbon: View > Palettes > External References Palette **Command:** IMAGE
Toolbar: Reference > External References **Menu Bar:** Insert > External References

Choose the **External References Palette** button from the **Palettes** panel; the **EXTERNAL REFERENCES** palette will be displayed. If the image has not been inserted earlier, right-click in the **File References** area and choose the **Attach Image** option from the shortcut menu. You can view the image information either as a list or as a tree view by choosing the respective buttons located at the right corner of the **File References** head of the **EXTERNAL REFERENCES** palette. The **List View** displays the names, loading status, size, date last modified on, and search path of all images in the drawing. The **Tree View** displays images in hierarchal structure. These images show its nesting levels within blocks and Xrefs. The **Tree View** does not display the status, size, or any other information about an image file. You can rename an image file in this dialog box.

EDITING RASTER IMAGE FILES

Raster image files can be easily edited. The different methods to edit a raster image files are discussed next.

Clipping Raster Images

Ribbon: Insert > Reference > Clip **Toolbar:** Reference > Clip Image
Menu Bar: Modify > Clip > Image **Command:** IMAGECLIP

The **Clip** tool is used to clip the boundaries of images and to provide a desired shape to them. To invoke this tool, select the image to be clipped; the **Image** tab will be added to the **Ribbon**. Choose the **Create Clipping Boundary** tool; you will be prompted to specify the boundary. Specify a rectangular or polygonal boundary; the image will be clipped at the specified boundary.

Adjusting Raster Images

Ribbon: Insert > Reference > Adjust **Menu Bar:** Modify > Object > Image > Adjust
Toolbar: Reference > Adjust Image **Command:** IMAGEADJUST/ADJUST

The **Adjust** tool allows you to adjust the brightness, contrast, and fade of a raster image. On selecting an attached image, the **Image** tab will be added to the **Ribbon**. Adjust the brightness, contrast, and fade using the corresponding option in the **Adjust** panel of the **Image** tab.

Modifying the Image Quality

Menu Bar: Modify > Object > Image > Quality **Command:** IMAGEQUALITY
Toolbar: Reference > Image Quality

The **Quality** tool allows you to control the quality of the image that affects the display performance. To control the quality, choose the **Quality** tool from **Modify > Object > Image** in the Menu bar; you will be prompted to specify the quality of image. A high-quality image takes a longer time to display. When you change the quality, the display changes immediately without causing a **REGEN**. The images are always plotted using a high-quality display. Although draft quality images appear grainier, they are displayed more quickly.

Modifying the Transparency of an Image

Toolbar: Reference > Image Transparency **Command:** TRANSPARENCY
Menu Bar: Modify > Object > Image > Transparency

When you attach an image with transparent background to a drawing, the transparent background turns opaque. To control the transparency of the attached image, select it; the **Image** tab will be added to the **Ribbon**. Choose the **Background Transparency** button from the **Image** tab to turn on the transparency of the image; the background of the image will become transparent. The **TRANSPARENCY** command can also be used to control the transparency of an image.

Controlling the Display of Image Frames

Menu Bar: Modify > Object > Image > Frame **Command:** IMAGEFRAME
Toolbar: Reference > Image Frame

The **FRAME** tool is used to turn the image boundary on or off. This tool can be invoked from **Modify > Object > Image** menu. If the image boundary is off, the image cannot be selected with the pointing device and therefore cannot be accidentally moved or modified.

DWG Convert

Application Menu: Save As > DWG Convert **Command:** DWGCONVERT
Menu Bar: File > DWG Convert

The **DWG Convert** tool is used to convert a single or batch of the AutoCAD drawing files into any older version of AutoCAD drawing format. You can do the conversion by following the procedure given next.

Invoke the **DWG Convert** dialog box by choosing the **DWG Convert** tool from the **File** menu; the **DWG Convert** dialog box will be displayed. Choose the **Add file** button from the bottom of the **Files Tree** area; the **Select File** dialog box will be displayed. Next, browse to the required location, select the required file/files, and then choose the **Open** button; the selected files will be displayed in tree form in the **Files Tree** area of the **DWG Convert** dialog box. Save the file list in the desired folder.

EXERCISE

Exercise 1

In this exercise, you will create a cup and a plate, refer to Figure 21-1 for dimensions. Assume the missing dimensions. Below the cup and plate, enter the following text in MS Word: **These objects are drawn in AutoCAD**. Then using OLE, paste the text into the current drawing.

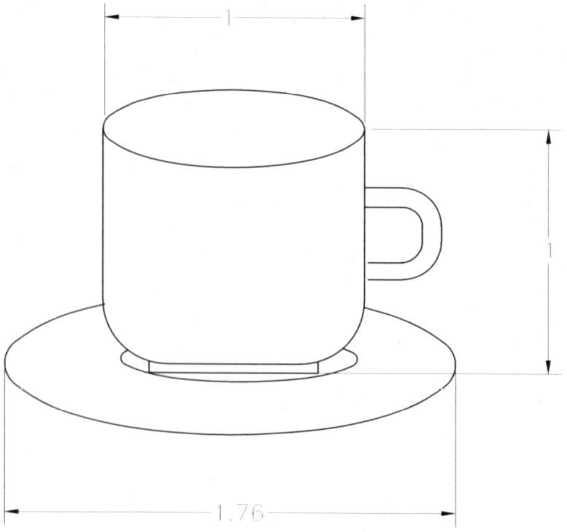

Figure 21-1 *Dimensions for Exercise 1*

Chapter 22

Isometric Drawings

Learning Objectives

After completing this chapter, you will be able to:

- *Understand isometric drawings, isometric axes, and isometric planes*
- *Set isometric grid and snap*
- *Draw isometric circles in different isoplanes*
- *Dimension isometric objects*
- *Write text in isometric styles*

ISOMETRIC DRAWINGS

Isometric drawings are generally used to help visualize the shape of an object. For example, if you are given the orthographic views of an object, as shown in Figure 22-1, it takes time to put information together to visualize the shape. However, if an isometric drawing is given, as shown in Figure 22-2, it is much easier to understand the shape of the object. Thus, isometric drawings are widely used in industry to help in understanding products and their features.

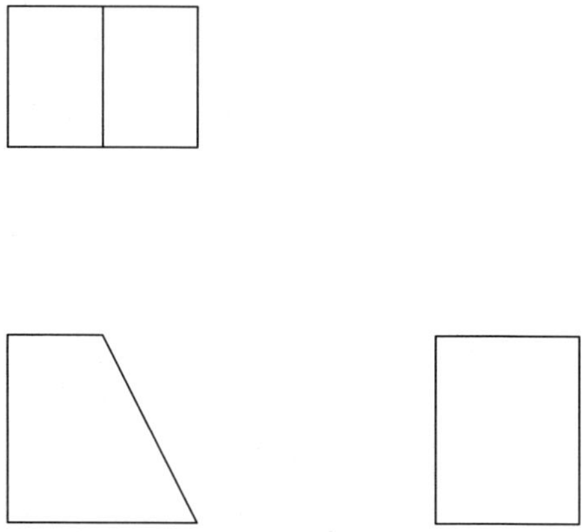

Figure 22-1 Orthographic views of an object

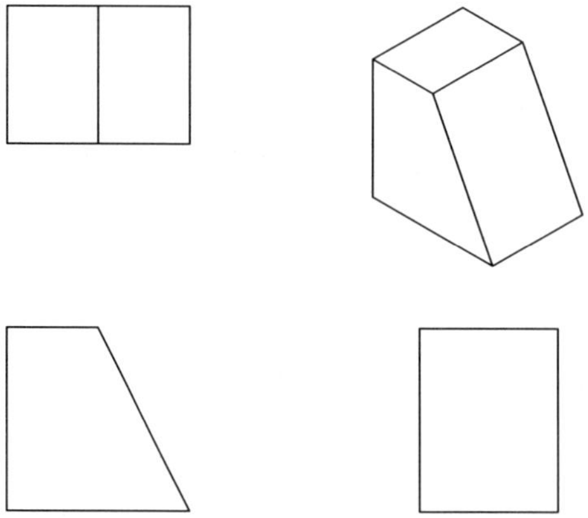

Figure 22-2 Orthographic views with an isometric view

An isometric drawing should not be confused with a three-dimensional (3D) drawing. An isometric drawing is just a two-dimensional (2D) representation of a 3D drawing on a 2D plane. A 3D drawing is the 3D model of an object on the X, Y, and Z axes. In other words, an isometric drawing is a 3D drawing on a 2D plane, whereas a 3D drawing is a true 3D model of the object. The model can be rotated and viewed from any direction. A 3D model can be a wireframe model, surface model, or solid model.

ISOMETRIC PROJECTIONS

The word "isometric" means equal measurement. The angle between any of the two principal axes of an isometric drawing is 120 degrees, refer to Figure 22-3. An isometric view is obtained by rotating the object by 45 degree angle around the imaginary vertical axis, and then tilting the object forward through a 35°16' angle. If you project the points and edges on the front plane, the projected length of the edges will be approximately 82 percent (isometric length/actual length = 9/11), which is shorter than the actual length of the edges. However, isometric drawings are always drawn to a full scale because their purpose is to help the user visualize the shape of the object. Isometric drawings are not meant to describe the actual size of the object. The actual dimensions, tolerances, and feature symbols must be shown in the orthographic views. Also, you should avoid showing any hidden lines in the isometric drawings, unless they show an important feature of the object or help in understanding its shape.

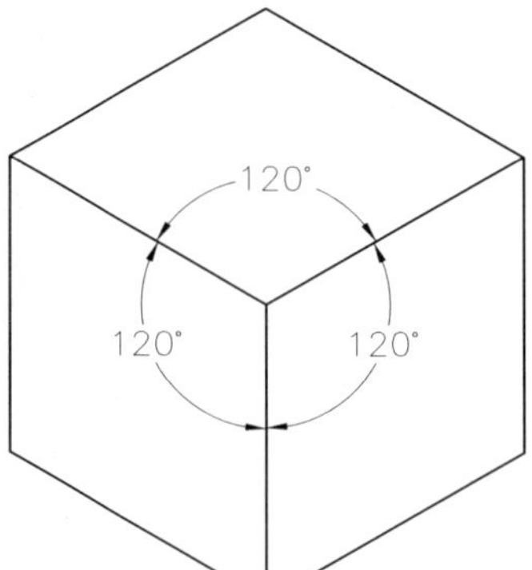

Figure 22-3 Principal axes of an isometric drawing

ISOMETRIC AXES AND PLANES

Isometric drawings have three axes: right horizontal axis (P0,P1), vertical axis (P0,P2), and left horizontal axis (P0,P3). The two horizontal axes are inclined at 30 degrees to the horizontal or X axis (X1,X2). The vertical axis is at 90 degrees, as shown in Figure 22-4.

When you draw an isometric drawing, the horizontal object lines are drawn along or parallel to the horizontal axis. Similarly, the vertical lines are drawn along or parallel to the vertical axis. For example, to make an isometric drawing of a rectangular block, the vertical edges of the block are drawn parallel to the vertical axis. The horizontal edges on the right side of the block are drawn parallel to the right horizontal axis (P0,P1), and the horizontal edges on the left side of the block are drawn parallel to the left horizontal axis (P0,P3). It is important to remember that the angles do not appear true in isometric drawings. Therefore, the edges or surfaces that are at an angle are drawn by locating their endpoints. The lines that are parallel to the isometric axes are called isometric lines. The lines that are not parallel to the isometric axes are called non isometric lines.

Similarly, the planes can be isometric planes or non isometric planes.

Isometric drawings have three principal planes, namely isoplane right, isoplane top, and isoplane left, as shown in Figure 22-5. The isoplane right (P0,P4,P10,P6) is defined by the vertical axis and the right horizontal axis. The isoplane top (P6,P10,P9,P7) is defined by the right and left horizontal axes. Similarly, the isoplane left (P0,P6,P7,P8) is defined by the vertical axis and the left horizontal axis.

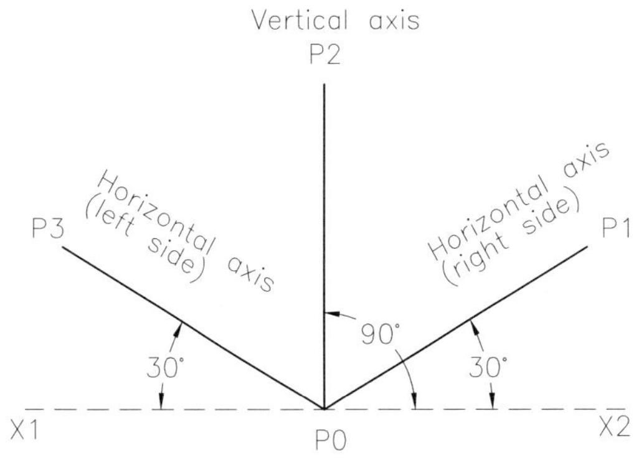

Figure 22-4 *Isometric axes*

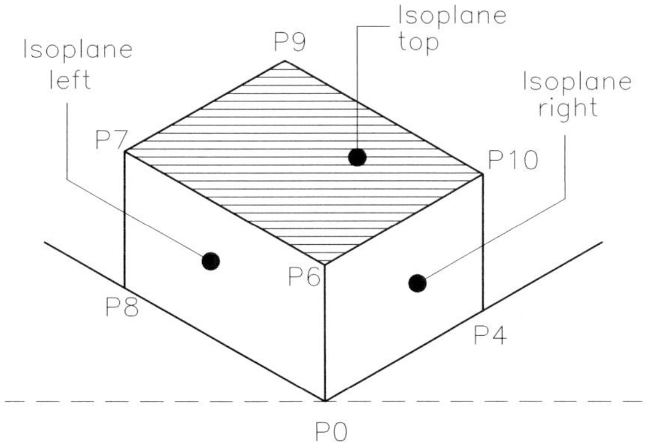

Figure 22-5 *Isometric planes*

SETTING THE ISOMETRIC GRID AND SNAP

You can use the **SNAP** command to set the isometric grid and snap. The isometric grid lines are displayed at 30 degree angle to the horizontal axis. Also, the distance between the grid lines is determined by the vertical spacing which can be specified by using the **GRID** or **SNAP** command. The grid lines coincide with three isometric axes which make it easier to create isometric drawings. The following command sequence and Figure 22-6 illustrate the use of the **SNAP** command to set the isometric grid and snap of 0.5 unit:

Command: **SNAP**
Specify snap spacing or [ON/OFF/Aspect/Legacy/Style/Type] <0.5000>: **S**
Enter snap grid style [Standard/Isometric] <S>: **I**
Specify vertical spacing <0.5000>: *Enter a new snap distance.*

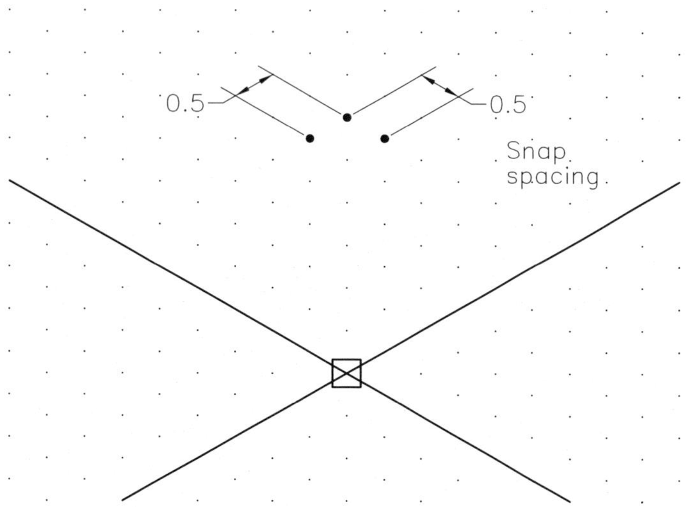

Figure 22-6 *Setting the isometric grid and snap in dotted grid*

Note

*1. When you use the **SNAP** command to set the isometric grid, the grid lines may not be displayed. To display the grid lines, turn the grid on by choosing the **Grid Display** button from the Status Bar or press F7.*

2. You cannot set the aspect ratio for the isometric grid. Therefore, the spacing between the isometric grid lines will be the same.

You can also set the isometric grid and snap by using the **Drafting Settings** dialog box shown in Figure 22-7. You can invoke this dialog box by right-clicking on **SNAPMODE**, **GRIDMODE**, **Polar Tracking**, **Object Snap**, **3D Object Snap**, **Object Snap Tracking**, **Dynamic Input**, **Quick Properties**, or **Selection Cycling** button available in the Status Bar and then choosing **Settings** from the shortcut menu displayed. You can also invoke this dialog box by entering **DSETTINGS** at the Command prompt.

The isometric snap and grid functions can be turned on/off by choosing the **Grid On (F7)** check box located in the **Snap and Grid** tab of the **Drafting Settings** dialog box. The **Snap and Grid** tab also contains the radio buttons to set the snap type and style. To display the grid on the screen, make sure the grid is turned on.

When you set the isometric grid, the display of crosshairs also changes. The crosshairs are displayed at an isometric angle and their orientation depends on current isoplane. You can toggle between isoplane right, isoplane left, and isoplane top by pressing the CTRL and E keys (CTRL+E) simultaneously or by using the function key, F5. You can also toggle among different isoplanes by entering the **ISOPLANE** command at the Command prompt:

Command: **ISOPLANE**
Enter isometric plane setting [Left/Top/Right] <Top>: **T**
Current Isoplane: **Top**

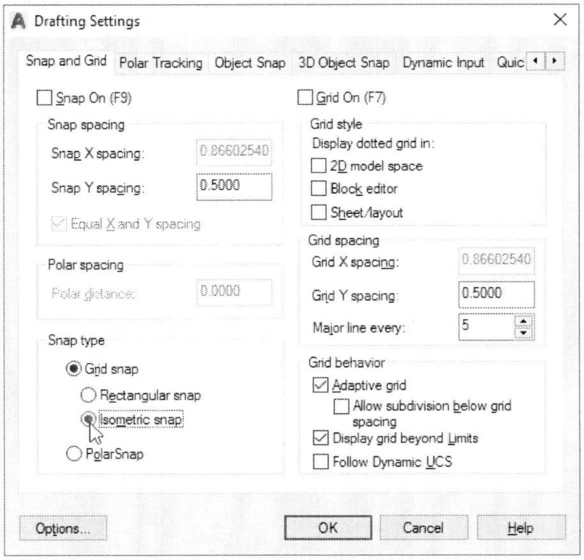

Figure 22-7 *The **Drafting Settings** dialog box*

The Ortho mode is useful while drawing in the Isometric mode. In the Isometric mode, Ortho aligns with the axes of the current isoplane.

Tutorial 1 — *Isometric Drawing*

In this tutorial, you will create the isometric drawing shown in Figure 22-8.

1. Use the **SNAP** command to set the isometric grid and snap. The snap value is 0.5 unit.

   Command: **SNAP**
   Specify snap spacing or [ON/OFF/Aspect/Legacy/Style/Type] <0.5000>: **S**
   Enter snap grid style [Standard/Isometric] <S>: **I**
   Specify vertical spacing <0.5000>: **0.5** *(or press ENTER)*

2. Change the isoplane to the isoplane left by pressing the F5 key. Choose the **Line** tool and draw lines between the points P1, P2, P3, P4, and P1, as shown in Figure 22-9.

Tip
*You can increase the size of the crosshairs using the **Crosshair size** slider bar in the **Display** tab of the **Options** dialog box.*

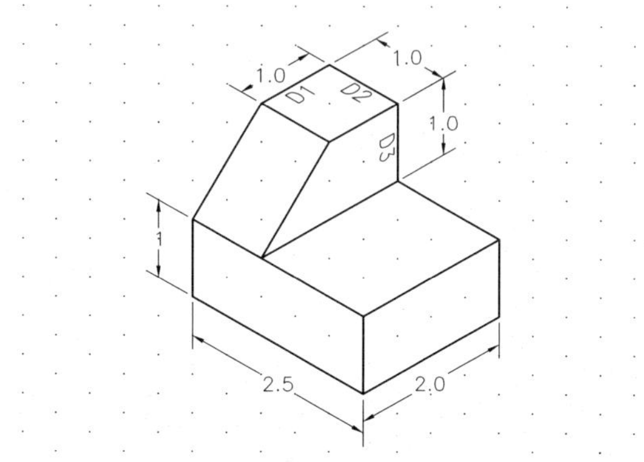

Figure 22-8 *Isometric drawing for Tutorial 1*

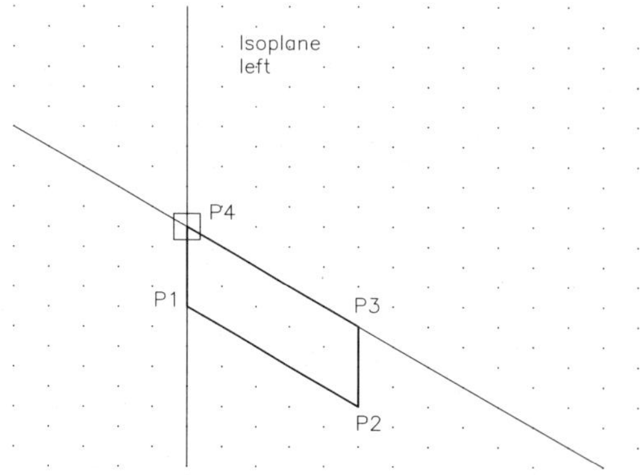

Figure 22-9 *Drawing the bottom left face*

3. Change the isoplane to the isoplane right by pressing the F5 key. Invoke the **Line** tool and draw the lines, as shown in Figure 22-10.

4. Change the isoplane to the isoplane top by pressing the F5 key. Invoke the **Line** tool and draw the lines, refer to Figure 22-11.

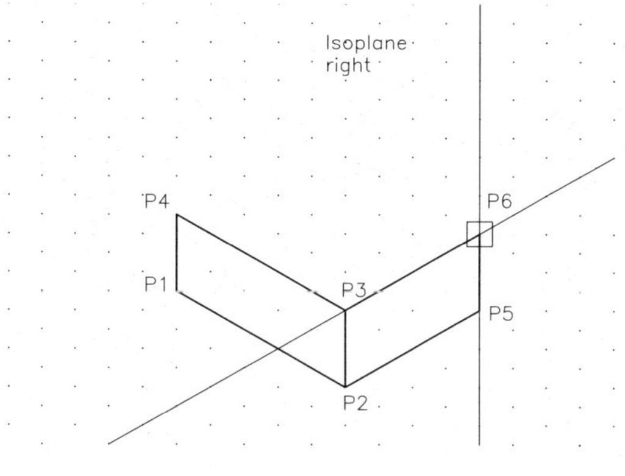

Figure 22-10 *Drawing the bottom right face*

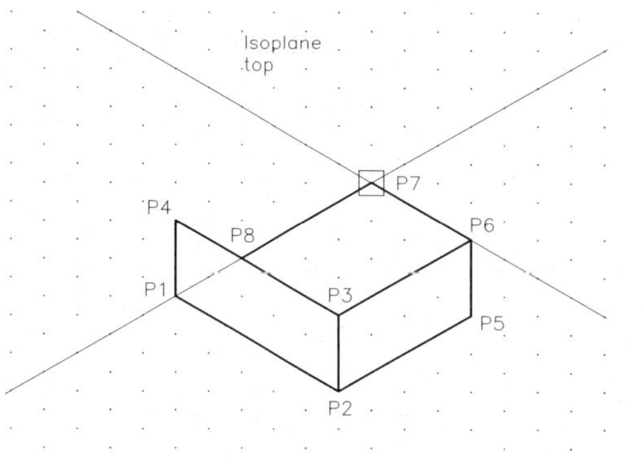

Figure 22-11 *Drawing the top face*

5. Similarly, draw the remaining lines, refer to Figure 22-12.

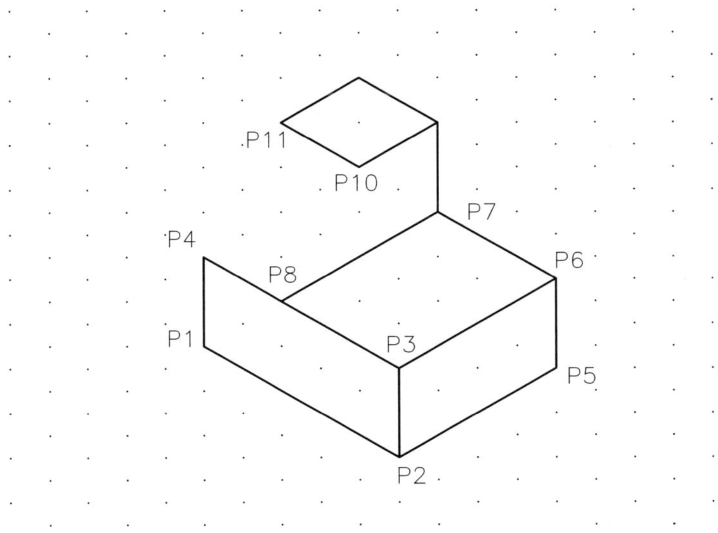

Figure 22-12 *Drawing the remaining lines*

6. The front left end of the object is tapered at an angle. In isometric drawings, oblique surfaces (surfaces at an angle to the isometric axis) cannot be drawn like other lines. Make sure that the Endpoint Object Snap is on, and then locate the endpoints of the lines that define the oblique surface. Next, draw the lines between those points. To complete the drawing shown in Figure 22-8, draw a line from P10 to P8 and from P11 to P4, refer to Figure 22-13.

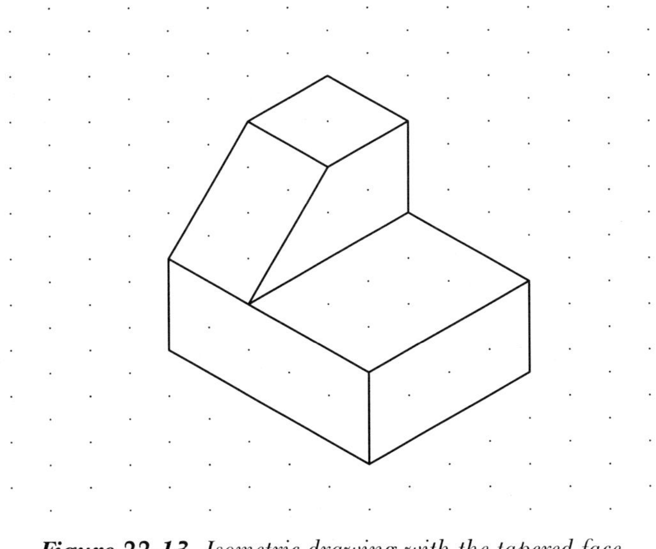

Figure 22-13 *Isometric drawing with the tapered face*

DRAWING ISOMETRIC CIRCLES

The isometric circles are drawn by using the tools available in the **Ellipse** drop-down and then selecting the **Isocircle** option. To draw an isometric circle, choose the **Axis, End**, or **Elliptical Arc** tool from the **Ellipse** drop-down. As soon as you select any of these options, the **Isocircle** option will be available in the Command prompt. Select the **Isocircle** option from the Command prompt; you will be prompted to specify the center of isocircle. Specify the center of the isocircle.

On specifying the center, you will be prompted to specify the radius of the isocircle. Once you specify the radius and press enter, isocircle will be created.

Note that you must have the Isometric Snap on while using the Ellipse tools to display the **Isocircle** option. If the isometric snap is not ON, you cannot draw an isometric circle. Before entering the radius or diameter of the isometric circle, you must make sure that you are in the required isoplane.

For example, to draw a circle in the right isoplane, you must toggle through the isoplanes until the required isoplane (right isoplane) is displayed. You can also set the required isoplane as the current plane before choosing the **Ellipse** tool. The crosshairs and the shape of the isometric circle will automatically change as you toggle through different isoplanes. As you enter the radius or diameter of the circle, AutoCAD draws the isometric circle in the selected plane, refer to Figure 22-14.

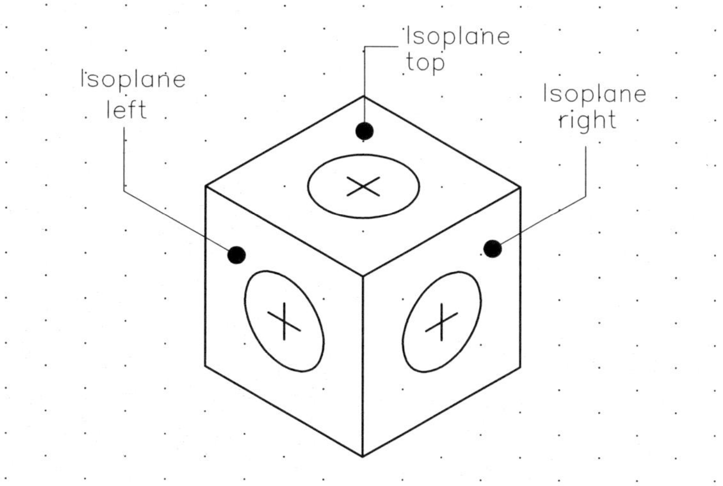

Figure 22-14 *Isometric circles drawn*

Tutorial 2 — Dimensioning

The following tutorial illustrates the process involved in dimensioning an isometric drawing. In this tutorial, you will dimension the isometric drawing created in Tutorial 1.

1. Dimension the drawing given in Tutorial 1, as shown in Figure 22-15. You can use the aligned or linear dimensions to dimension the drawing. Remember that when you select the points, you must use the **Intersection** or **Endpoint** object snap to snap the endpoints of the object you are dimensioning. AutoCAD automatically leaves a gap between the object line and the extension line, as specified by the **DIMGAP** variable.

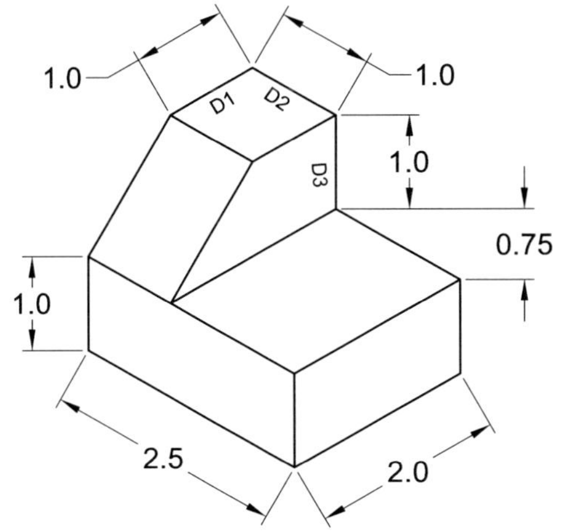

Figure 22-15 The dimensioned isometric drawing before using the **Oblique** tool

2. The next step is to edit the dimensions. You can choose the **Oblique** tool from the **Dimensions** panel of the **Annotate** tab. After selecting the dimension that you want to edit, you are prompted to enter the oblique angle. The oblique angle is determined by the angle that the extension line of the isometric dimension makes with the positive *X* axis. The following prompt sequence is displayed when you invoke this option from the **Ribbon**:

   Select objects: *Select the dimension (D1).*
   Select objects: *Press ENTER.*
   Enter obliquing angle (Press ENTER for none): **150** Enter

For example, the extension line of the dimension labeled D1 makes a 150 degree angle with the positive X axis, refer to Figure 22-16(a), therefore, the oblique angle is 150 degrees. Similarly, the extension lines of the dimension labeled D2 and D3 make a 30 degree angle with the positive X axis, refer to Figure 22-16(b) and (c), therefore, the oblique angle is 30 degrees. After you edit all dimensions, the drawing should appear as shown in Figure 22-17.

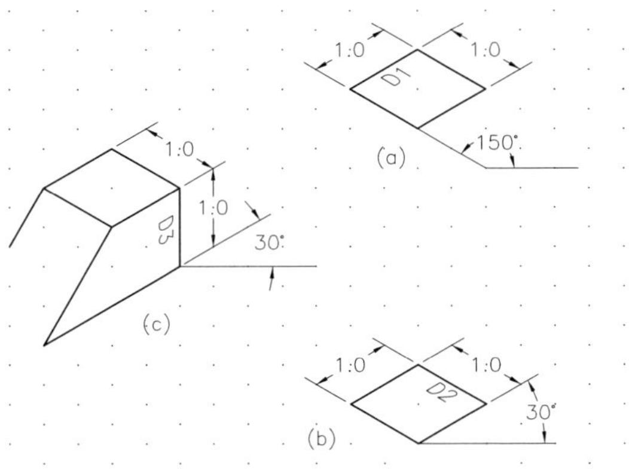

Figure 22-16 Determining the oblique angle

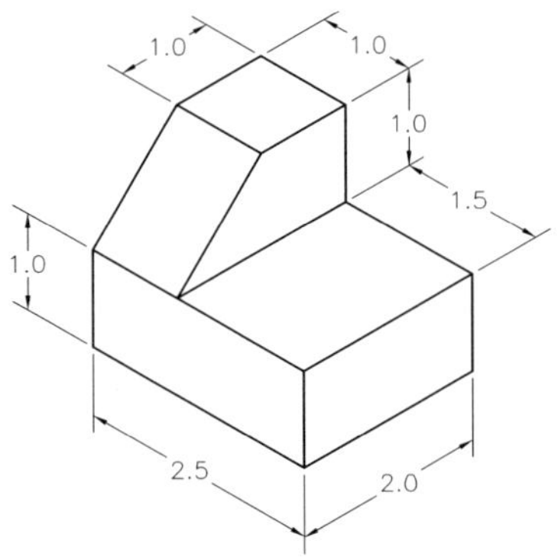

Figure 22-17 Object with isometric dimensions

EXERCISES

EXERCISES 1 through 4

Draw the following isometric drawings, refer to Figures 22-18 through 22-21. The dimensions can be determined by counting the number of grid lines. The distance between the isometric grid lines is assumed to be 0.5 unit. Dimension the drawing, as shown in Figure 22-21.

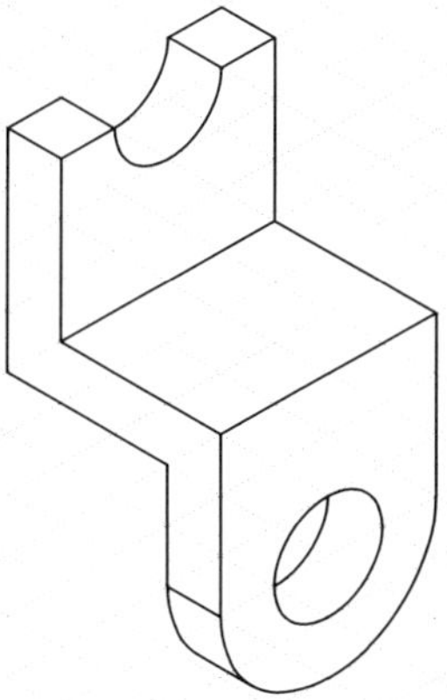

Figure 22-18 *Drawing for Exercise 1*

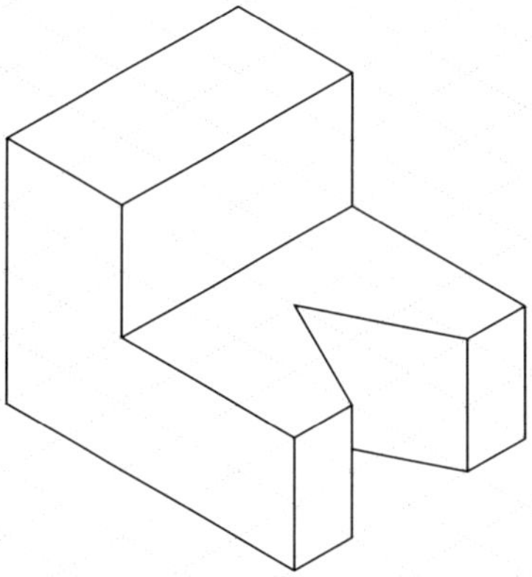

Figure 22-19 *Drawing for Exercise 2*

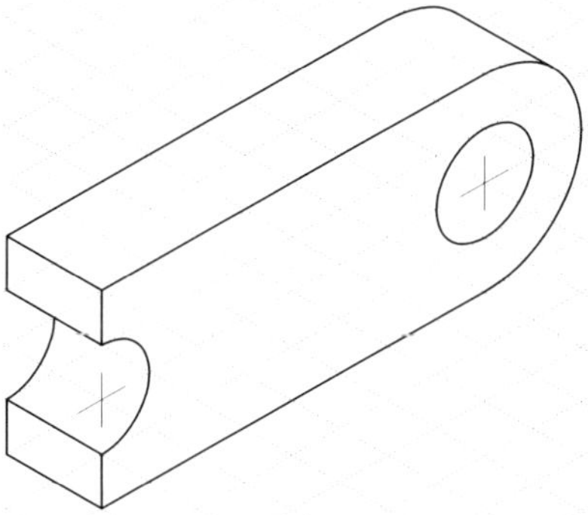

Figure 22-20 *Drawing for Exercise 3*

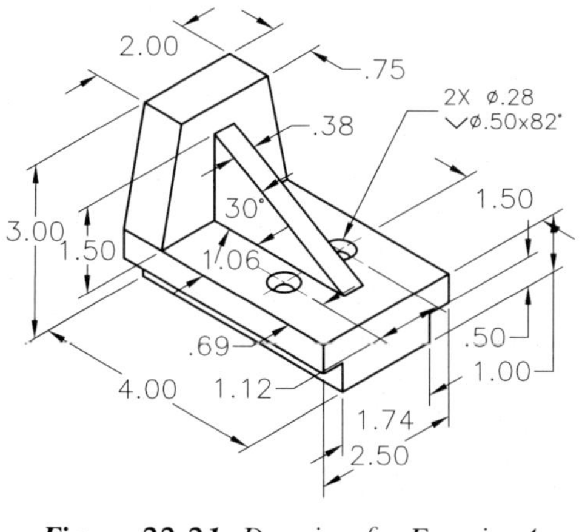

Figure 22-21 *Drawing for Exercise 4*

Index

Symbols

A

B

C

D

E

F

G

H

I

J

L

M

U

V

W

X

Made in United States
North Haven, CT
05 November 2026

21768329R10122